Worked Examples in Quantitative Metallography

Dedication

This book is dedicated to the memory of Jack Woodhead, lecturer and then senior lecturer in the department of Metallurgy, University of Sheffield from 1946 to 1983. He taught one of the authors, and instilled his conviction that microstructures must be specified numerically. The methods of quantifying microstructures illustrated in this book build on examples that have been used in teaching the subject at Sheffield and at CEIT, San Sebastian in courses that aimed to continue the tradition started by Jack, which has had a major influence on our research activities and on those of many others he taught.

Rebecca Higginson
Mike Sellars

Worked Examples in Quantitative Metallography

R. L. HIGGINSON
and
C. M. SELLARS

MANEY

FOR THE INSTITUTE OF MATERIALS, MINERALS AND MINING

B0788
First published for IOM[3] in 2003 by
Maney Publishing
1 Carlton House Terrace
London SW1Y 5DB

ISBN 1-902653-80-7

Typeset in India by Emptek Inc.
Printed and bound in the UK by the Charlesworth Group

Contents

Acknowledgements

The authors are grateful to Dr John Whiteman for reading, commenting on and checking earlier versions of the text. They are also grateful to a number of their colleagues, from whose work the examples have been taken, and particularly to Dr Qiang Zhu for Example 4.3.2.

1. Introduction

1.1 OBJECTIVES

In physical metallurgy it is frequently necessary to obtain quantitative measurements of microstructural features to compare experimental observations with theoretical predictions. These may relate to the kinetics of processes such as grain growth, phase transformations or particle coarsening, or to the development of mechanical properties such as strength and toughness. Quantitative metallography, or stereology, is concerned with the measurement of microstructural features such as grain size, and the size and spatial distribution of second phase particles from observations made on sections by optical or scanning electron microscopy, and on replicas or thin foils by transmission electron microscopy. In all cases only a small sample section or thin slice of material is observed in order to derive the microstructural characteristics in the bulk material. Stereology is therefore concerned with geometrical probability. The mathematics behind the analysis of experimental data has been developed over a long period from the middle of the nineteenth century.

The principles are described in a number of standard texts, for example, De Hoff and Rhines (1968), Underwood (1970), Pickering (1975). The aim of this book is to show examples of the application of these principles to calculate mean values and the confidence intervals for a number of important microstructural features. Knowledge of the accuracy of the experimental values of the microstructural parameters is essential for their valid interpretation and application. It is assumed that the reader is familiar with the application of statistical methods to data analysis, but the essentials are summarised in the Appendix. Considering the effort involved in making quantitative metallographic measurements, the statistical uncertainties of the answers are disappointingly wide unless many measurements are made. Effort versus accuracy must therefore be taken into account in experimental planning, as described in Chapter 2. The tedium of making measurements 'by hand' is being increasingly removed by sophisticated image processing and image analysis software, but it is usually necessary to calibrate the systems for the particular alloy, the thermomechanically processed condition, and the metallographic preparation procedure of interest to ensure that only the appropriate microstructural features are being examined. The examples in this book provide a basis for carrying out such calibrations and for understanding the measurements made by automated systems, as well as enabling valuable measurements to be made using standard metallographic equipment.

1.2 SYSTEMATIC ERRORS

In addition to the statistical uncertainties, which are considered in each chapter, there are a number of potential sources of systematic error that can arise from preparation of the specimen or observation area, or from the metallographic technique.

1.2.1 Specimen Preparation

It is self-evident that the sample to be examined should be representative of the bulk for which the microstructure is of interest, but this is not always simple, e.g. for castings or forgings in which the structures are heterogeneous. Careful consideration needs to be given to the purpose of making the microstructural measurements in order to choose the best sampling position to be examined. The size of the specimen taken must then be sufficiently large to ensure that no spurious effects are introduced from the surfaces if subsequent heat treatment or thermomechanical processing is given before the microstructure is examined. After such treatment, it is good practice to section the specimen again either centrally or well away from an original surface before preparing the section for metallographic examination. Qualitative observation then enables the depth of any surface affected zone, such as decarburisation of steel, to be assessed either for avoidance, or for quantification in subsequent measurements.

1.2.2 Depth of Etching

The mathematics for analysis of observations on a sectioned surface assumes that it is perfectly plane. However, in practice it is usually necessary to etch the surface so that the microstructure can be observed. This makes the less reactive phase or inert particles become proud of the surface and leads to systematic overestimation of the volume fraction (V_V) from observations on the surface for the reason illustrated in Fig. 1.1.

Particles that are sectioned below their maximum 'diameter' are observed at their correct size on the plane of polish, whereas particles sectioned above their maximum 'diameter' are observed to have a larger size, and etching may reveal some particles not sectioned by the plane of polish. This leads to a larger observed volume fraction (V_v^1). Cahn and Nutting (1959) showed that the true volume fraction can be obtained as

$$V_v = V_v^1 - \frac{S_v t}{4}$$

(1.1)

where t is the depth of etching, which is typically a fraction of a micrometer, and S_V is the surface area of the particles per unit volume of matrix, which can be measured as discussed in Chapter 4.

Etching depth in a microstructure containing inert second phase particles may be determined by sputtering the surface with gold at a known angle of inclination to the

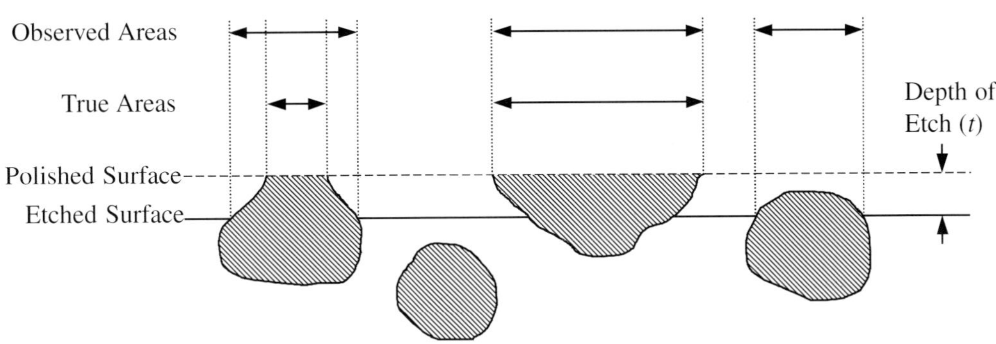

Fig. 1.1 Diagrammatic representation of sectioning errors (after Pickering (1975)).

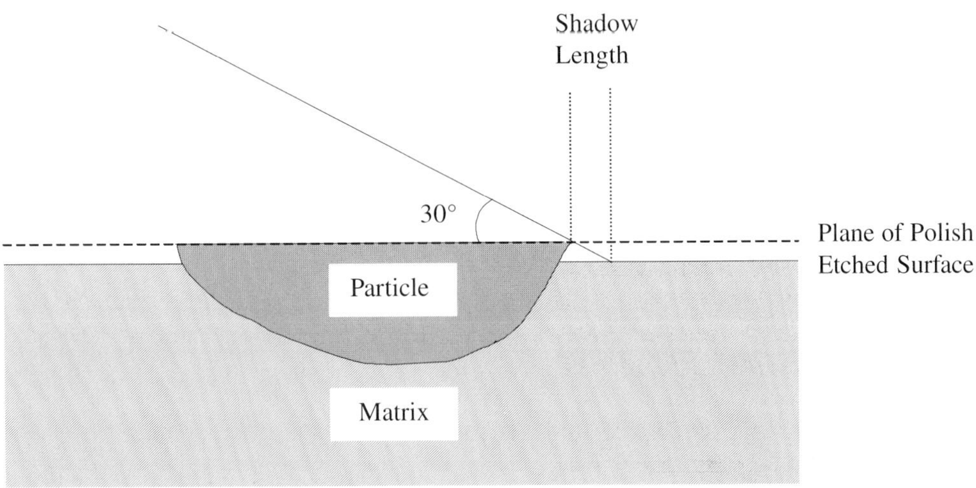

Fig. 1.2 Measurement of depth of etching.

surface (say 30°). This leads to 'shadows' on the matrix cast by the particles protruding from the etched surface. The maximum shadow lengths arise from particles sectioned below their maximum diameter, as illustrated in Fig. 1.2. The depth of etching is then simply calculated from the maximum shadow lengths.

EXAMPLE 1.2.2 – DEPTH OF ETCH

Shadowing the etched surface of a duplex stainless steel with gold at an angle of 30°
gave a maximum shadow length (measured in a scanning electron microscope) of 0.18
µm, cast by the austenite phase on the ferrite phase. Point counting, as described in
Chapter 3, gave a point fraction of austenite $\bar{P}_p = V_V^1 = 0.37 \pm 0.03$. Counting the number
of interphase boundaries per unit length as described in Chapter 4 leads to a value of
$S_v = 0.074$ µm^{-1}.

From the shadow length, the depth of etch

$$t = 0.18 \tan 30 = 0.104 \text{ µm}$$

Substitution into eqn (1.1) gives

$$V_V = 0.37 - \frac{0.074 \times 0.104}{4}$$

$$= 0.37 - 0.002$$

This systematic error is an order of magnitude smaller than the 95% confidence
interval and so can be neglected in this case. However, a somewhat refined microstructure,
i.e. higher S_V, a somewhat deeper etch, or a more accurate determination of V_V^1 would
mean that a correction for V_V may need to be made.

1.2.3 BOUNDARY THICKNESS

Etching may lead to the boundaries between phases appearing to be of finite thickness.
The boundaries of small particles may also appear to be of finite thickness in an optical
microscope because of the limit of resolution imposed by the wavelength of light (λ),
Fig. 1.3. Pickering (1975) estimated the overestimation of the volume fraction (δV_V) if
spherical particles are measured to the outer edge of the apparent boundary as

$$\delta V_V = \frac{2.4\lambda}{(N.A)} \frac{V_V}{D} \tag{1.2}$$

where $N.A.$ is the numerical aperture of the microscope and D is the particle diameter.

EXAMPLE 1.2.3 – BOUNDARY THICKNESS

Consider that measurements are made on a microscope with a numerical aperture of 1.3
using light of wavelength 0.52 µm, then substitution into eqn (1.2) gives the relative
errors in Table 1, if measurements are made to the outer edge of the particle images.
Clearly, when particles are only a few micrometers in size, the systematic error is serious.
This is best overcome by using a scanning electron microscope, for which λ is very

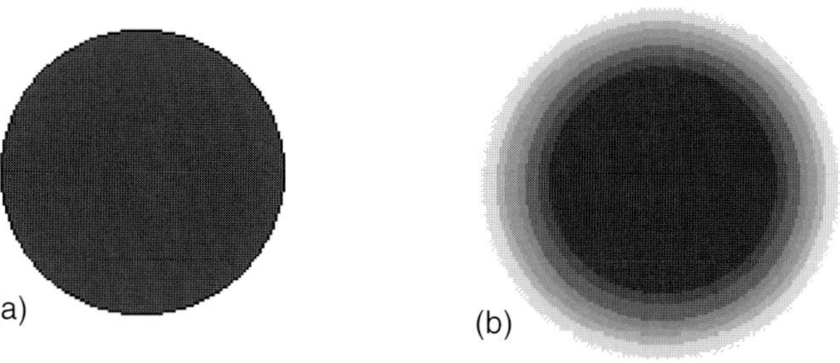

Fig. 1.3 Appearance of particle boundaries in the optical microscope: (a) large particle at low magnification and (b) small particle at high magnification.

Table 1 Boundary thickness error.

Particle Diameter µm	Relative Error $\delta\, V_v/V_v$ (%)
1000	0.09
100	0.94
10	9.4
1	94

much smaller, but in an optical microscope the errors can be minimised by measuring to the midpoint of the boundary image. This is effectively done in point counting if a count of 1 is made when a point is clearly within a particle, a count of 0 is made when a point is in the matrix and a count of 1/2 is made when a point is in the boundary region (c.f. Chapter 3).

1.2.4 SAMPLING

The mathematical basis of quantitative metallography assumes that microstructures are sampled randomly. For optical microscopy it is therefore best to make measurements directly in the microscope by traversing the prepared surface under the eyepiece or screen, so that the regions to be measured are 'sight unseen' before the measurement is made. For some optical metallographic measurements and for all electron metallographic measurements it is necessary to use photographs. Care must then be exercised to avoid introducing systematic errors by selecting the areas to be photographed.

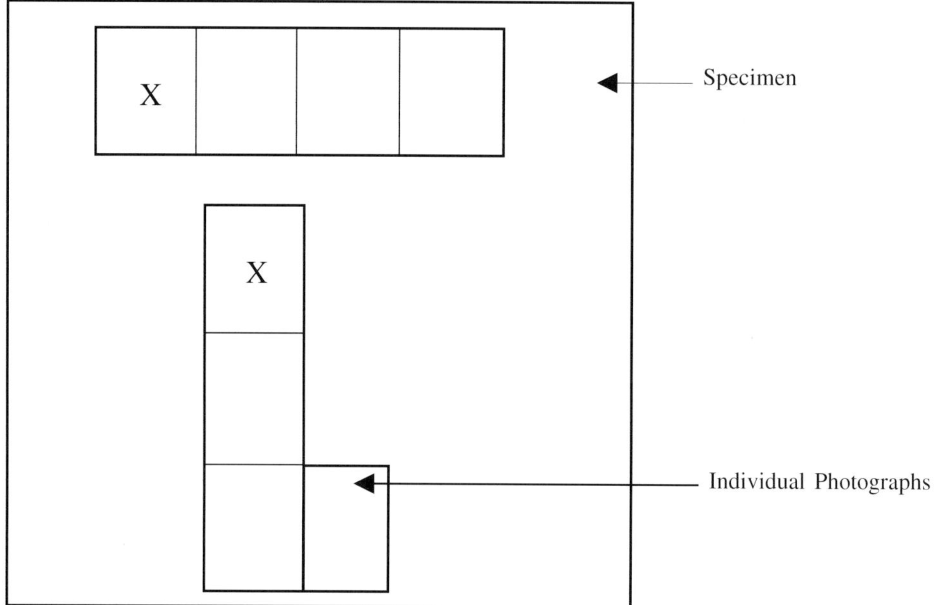

Fig. 1.4 Method of randomising the sample areas photographed.

In publications, micrographs are frequently captioned 'typical', but are generally selected to illustrate well some feature of the structure, and are sometimes simply the 'best' areas in terms of freedom from scratches or other preparation artefacts. A subjective selection has therefore been made, which is justifiable for qualitative illustration, but does not necessarily provide 'random' images for quantitative measurement. To do this, either a photograph must cover a sufficiently large number of features to be considered a random sample, or several photographs must be taken with most being of 'sight unseen' areas. This can be done as illustrated in Fig. 1.4, in which the photographs marked X might be selected for reproduction in a publication, but the others are simply adjacent areas found by traversing the specimen.

1.2.5 MEASUREMENT ERRORS

Systematic errors can also arise from experimental errors of measurement. Some common ones are:

- errors in calibrating magnification
- counting features that should not be counted, e.g. counting twin boundaries as well as grain boundaries in the determination of grain size.

- omitting to count features that should be counted, e.g. grain boundaries that are not clearly revealed by etching
- misidentification of phases in a duplex structure, e.g. distinguishing between recrystallised and unrecrystallised grains in partially recrystallised microstructures.

When quantitative metallographic measurements are made 'by hand' these types of errors can be avoided by experimental care and by accurate image interpretation, which comes with experience and understanding. For someone starting such measurements, it is a good idea to re-measure a sample already measured by an experienced colleague. If the sample means are close, within the 95% confidence limits, it provides some reassurance that no serious errors of recognition have been made.

These types of systematic errors can also occur in automatic image analysis, even though manufacturers provide image processing software to try to minimise them.

The recently introduces technique of Electron Back Scatter Diffraction (EBSD) using Scanning Electron Microscopy (SEM) to produce Orientation Imaging Micrographs (OIM) can reduce the possibility of such systematic errors, because the image contrast depends on the crystal structure and orientation of the microstructural features. Also, because the images are collected digitally, the technique facilitates automatic image analysis. It is therefore a powerful method of obtaining metallographic data previously found from optical micrographs, with the additional benefit of providing quantitative information about grain and sub-grain orientations. EBSD is being increasingly used as a research tool, with rapid development of the imaging interpretation software. This software uses the principles illustrated in the examples from optical microscopy, but it must be recognised that microstructural features, e.g. grain boundaries, recognised by misorientations across them may not always coincide with the boundaries revealed by etching for optical or scanning electron microscopic imaging. Thus care in interpretation and comparison of EBSD data with previously available metallographic data is required. This is still the subject of research and the methodology has yet to be standardised. EBSD images are therefore not specifically considered in the examples in this book.

2. Experimental Planning

As emphasised in Chapter 1, the accuracy of microstructural measurements is a key issue in experimental planning, once it has been decided which microstructural features are to be measured. It is clear from statistical analysis of data (see Appendix) that accuracy improves with the square root of the number of measurements made. However, the effort of making measurements increases linearly with their number. It is therefore usually necessary to arrive at a practicable compromise in planning experiments and to do this it is useful to be able to estimate the standard errors, or the confidence limits, to be expected before the experiments are carried out. Some simple examples are given in this Chapter.

2.1 VOLUME FRACTION

To determine volume fractions of phases, which are on a scale of micrometres (microns) to hundreds of micrometres, measurements are made on sections examined in the optical or scanning electron microscope, as discussed in Chapter 3.

2.1.1 POINT COUNTING

This method is based on counting the fraction of points (P_p), which fall in the phase of interest, in a random array of a total number of points (P). This could, for example, be pearlite in normalised steel, β phase in an $\alpha - \beta$ brass or recrystallised regions in a partially recrystallised structure. Point counting is generally the quickest and most statistically efficient way when volume fraction is to be determined by hand. The volume fraction.

$$V_V = \overline{P}_P \tag{2.1}$$

and the expected relative standard error in the value of V_V for a minor phase can be estimated from the relationship of Gladman and Woodhead (1960).

$$\left(\frac{S(V_V)}{V_V}\right)^2 = \frac{(1-V_V)}{PV_V} \approx \frac{(1-\overline{P}_P)}{P\overline{P}_P} \tag{2.2}$$

Clearly, the accuracy depends on the volume fraction expected, which can usually be estimated approximately from the phase diagram, or may vary in a series of experiments on transformation kinetics from values of 0 to 1. Note that when V_V reaches 0.5, the

Table 2.1 Relative errors of volume fraction by point counting.

(1)	(2)	(3)		
Relative Error, $\dfrac{S(V_V)}{V_V}$	Relative 95% C.L.	Number of Points P		
		$V_V = 0.01$	0.1	0.5
0.01	± 2%	990000	90000	10000
0.025	± 5%	158400	14400	1600
0.050	± 10%	39600	3600	400
0.100	± 20%	9900	900	100

minor phase becomes the major phase, so that by considering that V_V always refers to the minor phase in eqn (2.2), its value can range from 0 to 0.5. Substitution in eqn (2.2) leads to the values given in Table 2.1. For point counting by hand, even with special counting equipment on the microscope, 1000 points may be considered a practical upper limit. This severely restricts the relative accuracy achievable, particularly for low volume fractions.

In many practical situations the absolute accuracy may be of importance. eqn (2.2) can then by rearranged to give:

$$(S(V_V))^2 = \frac{V_V(1-V_V)}{P} \approx \frac{\overline{P}_P(1-\overline{P}_P)}{P} \tag{2.3}$$

It is notable that eqn (2.3) is symmetric about the value of $\overline{P}_p = 0.5$, which is significant when the point fraction of a specific phase is of interest, e.g. for study of phase transformations, $\overline{P}_p$ can range from 0 to 1.0.

EXAMPLE 2.1.1 – PHASE TRANSFORMATION

Applying eqn (2.3) to measurements of the kinetics of a transformation described by the form of equation for diffusion controlled transformation derived independently by Johnson and Mehl, (1939), Avrami, (1939), and Kolmogorov, (1937), and now referred to by the names of the separate authors, or as the JMAK equation:

$$X = V_V = 1 - \exp - 0.693\left(\frac{t}{t_{50}}\right)^k \tag{2.4}$$

Table 2.2 Confidence limits for phase transformation.

(1)	(2)	(3)	(4)
X	**95% Confidence Range of X**	**Range of log.ln(1/(1-X))**	**log t/t_{50}**
0.1	0.081 to 0.119	-1.073 to -0.987	-0.409
0.25	0.223 to 0.277	-0.598 to -0.489	-0.191
0.50	0.468 to 0.532	-0.200 to -0.120	0
0.75	0.723 to 0.777	0.108 to 0.176	0.151
0.90	0.881 to 0.919	0.328 to 0.400	0.261

where t_{50} is the time to 50% transformation and k is a constant. Considering that 1000 points are measured on samples with values of X ranging from 0.1–0.9 and substituting into eqn (2.3) leads to the values of 95% confidence limits:

$$95\%CL \approx 2S(V_V) \approx 2\left[\frac{X(1-X)}{1000}\right]^{1/2}$$

with the results shown as the range of X in Column 2 of Table 2.2. In order to determine the constants in eqn (2.4) it is usual to plot log.ln(1/(1–X)) versus log t to obtain the value of k from the slope and the value of t_{50} from the time when log.ln(1/(1–X)) = - 0.159. The range of values of the function is shown for each value of X in Column 3 of Table 2.2. In this example the time is calculated by assuming that $k = 2$ in eqn (2.4) and is normalised as t/t_{50}, to give the values of the logarithm in Column 4. From these values, graphs of the transformation kinetics are plotted in Fig. 2.1. From Fig. 2.1(a), it can be seen that the measurements from 1000 points lead to a value of k in the range 1.9 to 2.1 and t/t_{50} ranging from 0.95 to 1.05. The effect of these statistical uncertainties on the transformation curve is shown in Fig. 2.1(b).

COMMENTS

1. It can be seen that statistical uncertainties from measurements of 1000 points in Fig. 2.1(b) are small, and are probably less than random experimental errors in other variables affecting the value of X at a given time. Thus in practice measurement of 400 to 500 points may be satisfactory.
2. Considering the golden rule that for valid application of the statistics for random sampling no two adjacent traverse lines should cross the same region of minor phase and no two adjacent points should fall in the same region of minor phase means that the area of the section required for measurement increases with the scale of the structure. For example, with a matrix of points of spacing 50 µm arranged roughly in a square, 1000 points requires an area ~ 1.6 × 1.6 mm for examination.

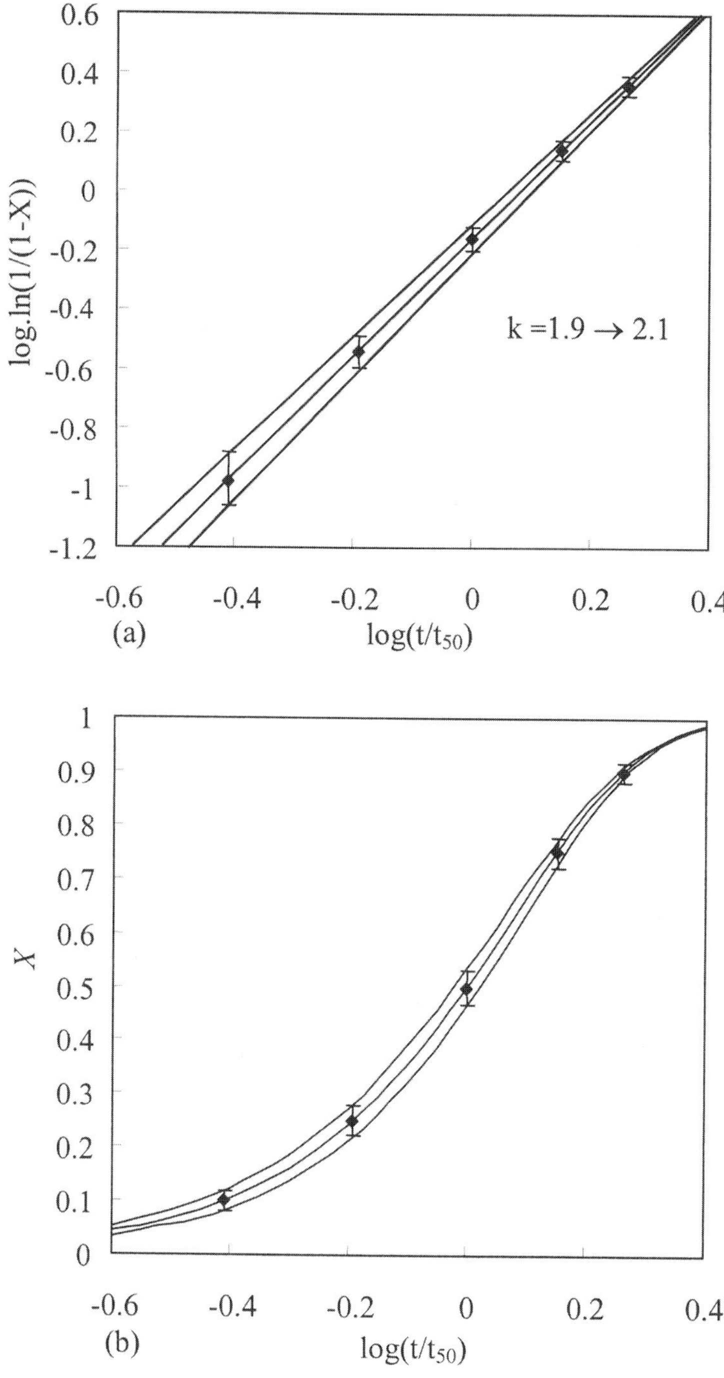

Fig. 2.1 (a) Influence of 95% confidence limits from measurements of 1000 points on the values of t_{50} and k in the JMAK equation and (b) the effect of the uncertainties on the transformation curve.

This would not normally be a problem, but for coarser structures consideration may need to be given to the specimen size and shape, particularly if rapid quenching is also required during experiments to achieve the microstructure of interest.

2.1.2 Areal Analysis

This method was the first one proposed for measurement of volume fraction (Delesse, 1848), but is extremely tedious to carry out by hand. However, automatic image analysis systems measure small increments of area, or pixels, and so obtain the area of each region of each phase in the section area and hence obtain the area fraction $\overline{A}_A$. The volume fraction

$$V_V = \overline{A}_A \tag{2.5}$$

and the expected relative standard error in the value of V_v can be estimated from the relationship derived by Hilliard and Cahn (1961):

$$\left(\frac{S(V_V)}{V_V}\right)^2 = \frac{1}{n}\left(1 + \left(\frac{\sigma_{(A)}}{\overline{A}}\right)^2\right) \tag{2.6}$$

where n is the number of areas of minor phase (α) measured, $\overline{A}$ is their average area and $\sigma_{(A)}$ is the standard deviation of the measured areas. For the most regular two phase structure comprising uniform spheres of α-phase, they show that $(\sigma_{(A)}/\overline{A})^2 = 0.2$. However, for two phase structures encountered in practice Woodhead (1980) suggests a likely value is $(\sigma_{(A)}/\overline{A})^2 = 0.5$. Substitution of this value in eqn (2.6) leads to

$$\frac{S(V_V)}{V_V} \approx \frac{1.25}{\sqrt{n}} \tag{2.7}$$

The relative standard error in this case is therefore independent of the volume fraction and leads to the number of areas to be measured given in Table 2.3.

Table 2.3 Relative errors of volume fraction by areal analysis.

(1)	(2)	(3)
Relative Error, $\dfrac{S(V_V)}{V_V}$	Relative 95% C.L.	Number of Areas, n
0.01	± 2%	15626
0.025	± 5%	2500
0.050	± 10%	625
0.100	± 20%	156

It can be seen that the numbers of areas to be measured are of the same order as given for point counting in Table 2.1. However, n should really be compared with the number of points falling in the α-phase, $P_a = V_V P$, indicating the greater statistical efficiency of point counting. Because areal analysis is carried out on automatic systems the number of measurements is a less critical factor, but the implications for the area of section required for measurement are more significant than for point counting, e.g. to achieve confidence limits of $\pm$ 10% for a structure which contains α-phase particles of about 25×25 µm requires measurement of 625 particles, i.e. a specimen area of $625 \times 0.025 \times 0.025 \times 10 \sim 2 \times 2$ mm when $V_V = 0.1$.

2.1.3 Lineal Analysis

This method is based on the fundamental relationship that volume fraction is equal to the mean line fraction of a phase determined along traverses on sections $(\overline{L}_L)$, i.e.

$$V_V = \overline{L}_L \tag{2.8}$$

The expected value of standard error of V_V can be estimated from the relationship derived by Gladman (1963):

$$\left(\frac{S(V_V)}{V_V} \right) = \frac{2}{n}(1 - \overline{L}_L)^2 \tag{2.9}$$

where n is the number of minor phase particles measured. Equation (2.9) leads to the numbers of particles to be measured shown in Table 2.4. It can be seen that the values of n are less than the values of P in Table 2.1, but again n should really be compared with the number of points in the α-phase, $P_a = V_V P$, which means that point counting is statistically more efficient. Nevertheless, lineal analysis provides a viable alternative method for low volume fractions of second phase, even when measurements are made 'by hand'.

Table 2.4 Relative errors of volume fraction by lineal analysis.

(1)	(2)	(3)		
Relative Error $\dfrac{S(V_V)}{V_V}$	**Relative** **95% CL**	**Number of Particles, n**		
		$V_V = 0.01$	0.1	0.5
0.01	±2%	19602	16200	5000
0.025	±5%	3136	2592	800
0.050	±10%	784	648	200
0.100	±20%	196	162	50

2.2 GRAIN SIZE

Measurement of grain size is considered in Chapter 4. The standard measures are made on sections, whereas grains are 3-dimensional solid crystals.

Even if the volume of a metal comprised grains of uniform size and shape, observations on a plane section would show a size variation from zero to some maximum value equal to the maximum tangent diameter of the grains in the volume. The frequency distributions of linear intercepts and of planar areas have been derived from this idealised structure, but for a real metal, in which there is a distribution of grain sizes in the volume, the observed frequency distributions are wider than ideal.

2.2.1 LINEAR INTERCEPT AND GRAIN AREA

From the experimental observations of Pereira da Silva (1966), Woodhead showed that the relative standard errors of the mean linear intercept grain size $(\bar{L})$ and the number of grains per unit length $(\bar{N}_L)$ are

$$\frac{S(\bar{L})}{\bar{L}} = \frac{S(\bar{N}_L)}{\bar{N}_l} \approx \frac{0.65}{\sqrt{n}} \tag{2.10}$$

where n is the number of grains counted.

For the mean grain area $(\bar{A})$ and the number of grains per unit area $(\bar{N}_A)$ the relative standard errors are

$$\frac{S(\bar{A})}{\bar{A}} = \frac{S(\bar{N}_A)}{\bar{N}_A} \approx \frac{1.03}{\sqrt{n}} \tag{2.11}$$

The numbers of grains that must be counted for a given accuracy are shown in Table 2.5.

Table 2.5 Accuracy of grain size measurement.

(1)	(2)	(3)	(4)
Relative Error	Relative 95% C.L.	Number of Intercepts for $\bar{L}$	Number of Grains for $\bar{A}$
0.01	± 2%	4225	10609
0.025	± 5%	676	1798
0.05	± 10%	169	424
0.10	± 20%	42	106

As for volume fraction, the specimen size needed depends on the accuracy required and the grain size of the material. For example, for a mean grain size of 50 µm, from Table 2.5, linear intercept measurement to obtain a confidence limit of ± 2.5 µm (i.e. ± 5%), requires a line length of 33.8 mm. In practice one would always measure along a number of shorter lines in order to obtain the confidence limits as well as the mean value of grain size. As discussed in Chapter 3, the largest grain in the section may well be about five times as large as the mean linear intercept, i.e. ~ 250 µm in diameter. Considering the golden rule that no two adjacent traverse lines should cross the same grain means that the lines should be at least 250 µm apart. In this case, 10 traverses would require a specimen area of at least 3.4 × 2.3 mm. Again, this would not normally be a problem, but for coarser grain sizes this criterion may be significant in the experimental design.

2.2.2 ASTM GRAIN SIZE NUMBER

ASTM values are frequently estimated by comparison of the microstructure with standard charts. This is a rapid method but liable to unknown subjective error. If truly quantitative values are required, they can be obtained from measurements of $\bar{N}_A$, see Chapter 4. The accuracy of ASTM grain size numbers (g) may then be determined simply from eqn (2.11), but, as g varies logarithmically with grain size, the absolute rather than the relative standard error is obtained as

$$S(\bar{g}) \approx \frac{1.49}{\sqrt{n}} \tag{2.12}$$

The number of grains that must be measured for a given accuracy is shown in Table 2.6.

Table 2.6 Accuracy of ASTM grain size numbers.

(1)	(2)	(3)
Standard Error	**95% C.L.**	**Number of Grains**
0.025	± 0.05	3552
0.05	± 0.10	883
0.10	± 0.20	221
0.25	± 0.50	36

2.3 PARTICLE SIZE

Size distributions of second phase particles vary so widely that it is only possible to provide a qualitative general guideline for measurements. Frequently size distributions are skewed to smaller sizes and are best measured as histograms with ten to fifteen size intervals. Usually several hundred particles must be measured to obtain the mean with reasonable accuracy. The observational technique depends on the mean and standard deviation of the size distribution, and on the purpose for which the results are to be used. For example, for particle size distributions with means of more that 10 μm, optical microscopy is generally preferred, but for means of less than 5 μm it is not normally satisfactory to use optical microscopy, because of the limited resolution. Scanning electron microscopy should be employed when the microstructures are to be related to strength, which depends on interparticle spacing. This may be dominated by the many small particles so that accurate measurement in this size range is essential. However, fatigue failure may be initiated at the largest particles in a distribution, with mean size of about 10 μm, such as inclusions in clean steels. In this case the distribution of large particles in the bulk steel can be estimated from the distribution of inclusion sizes in specimens above some threshold value of 3 to 4 μm, so that optical measurement is suitable (Shi et al., (1999)).

Scanning electron microscopy may be used for mean sizes down to about 0.5 μm but for smaller sizes it is necessary to use transmission electron microscopy. Extraction replicas enable relatively large numbers of particles to be measured more easily than by using thin foils, but it may not be possible to extract small particles, say less than 3 to 4 nm. A realistic lower limit of mean particle size for extraction replicas may therefore be about 25 nm for accurate measurement, with an upper limit of about 500 nm (0.5 μm) because of increasing difficulty of releasing replicas from the specimen surface. It should be noted (as discussed in Chapter 5) that extraction replicas are not suitable for determining numbers of particles per unit volume, or volume fraction.

For smaller particle sizes, thin foil microscopy is the preferred method for measurement, but it must be recognised that the volume of material examined in a thin foil micrograph is very small (typically less than 10^{-15} m^3), so that several micrographs, preferably from several thin foils may need to be examined. Unless the particle size is small, or volume fraction is high, this requirement places a practical upper limit of mean particle sizes for measurement of about 500 nm. If sufficient micrographs are examined for the sample to be considered random, then the number of particles per unit volume and volume fraction may be determined as well as particle size.

3. Volume Fraction from Planar Sections

Measurement of volume fraction (V_V) of second phase particles may be made from planar sections by optical or scanning electron microscopy because of the fundamental relationships

$$V_V = \overline{A}_A \qquad (3.1)$$

$$V_V = \overline{L}_L \qquad (3.2)$$

$$V_V = \overline{P}_P \qquad (3.3)$$

where $\overline{A}_A$ is the mean area fraction, $\overline{L}_L$ is the mean line fraction and $\overline{P}_P$ is the mean point fraction of second phase in a planar section. For making measurements 'by hand', point counting to obtain point fraction is generally the most efficient method, but either areal analysis or lineal analysis may be carried out using automatic image analysis systems.

3.1 POINT COUNTING

Fig. 3.1 shows a micrograph of a titanium microalloyed steel of composition 0.12% C, 1.35% Mn, 0.18% Si, 0.023% Al, 0.07% Ti, 0.14% N, <0.02% each of the other elements. The steel has been hot rolled to plate at a finishing temperature of 965°C and air cooled to room temperature.

The structure is colonies of unresolved pearlite (black) in a matrix of ferrite, in which the grain boundaries have also been etched as black lines. In point counting, care must be taken that ferrite grain boundaries are not mistakenly counted as pearlite colonies.

If the microstructure is nearly random, or at least is not regular, which is the usual situation in alloys, a regular array of points may be superimposed on the image and the number of points falling in the phase of interest is counted. This is usually a minor phase. The points may be superimposed in several ways:

EXAMPLE 3.1.1 – USING A GRATICULE

Method

A graticule of lines may be inserted in the eyepiece of an optical microscope or placed on the projection screen. For volume fractions of ~0.2 to 0.5 a graticule of 4 intersection points is suitable. The magnification must be chosen so that no two adjacent points fall

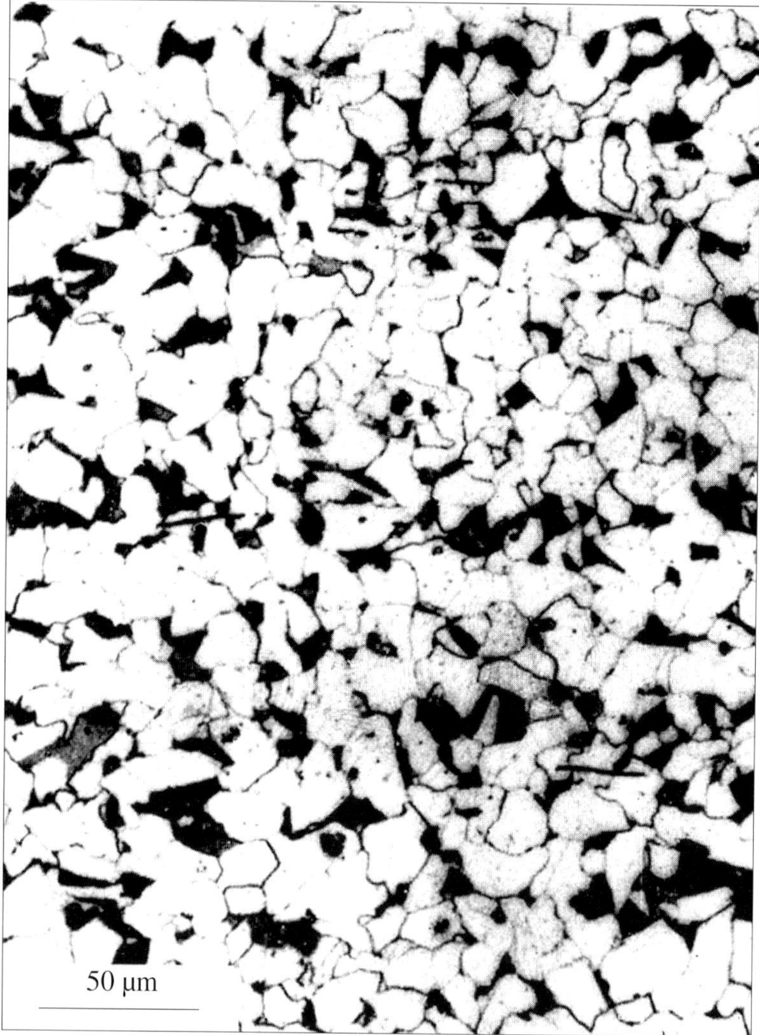

Fig. 3.1 Ferrite–pearlite structure in a microalloyed steel. Etched 2% Nital.

in the same area of minor phase (pearlite), as illustrated in Fig. 3.2. It is not necessary to know the magnification if only volume fraction is to be measured. In Fig. 3.2 the circle labelled (1) represents the first image position, showing one point in pearlite. This is noted and the specimen is then moved so that the graticule falls on a new field of view, labelled (2). Note that no point in the graticule should fall in the same area of pearlite as seen in the first image position. As long as this condition is met, the distance the specimen is moved is unimportant. The two points falling in pearlite are noted and the specimen is moved to position (3) etc. If a second traverse is made, such as shown by

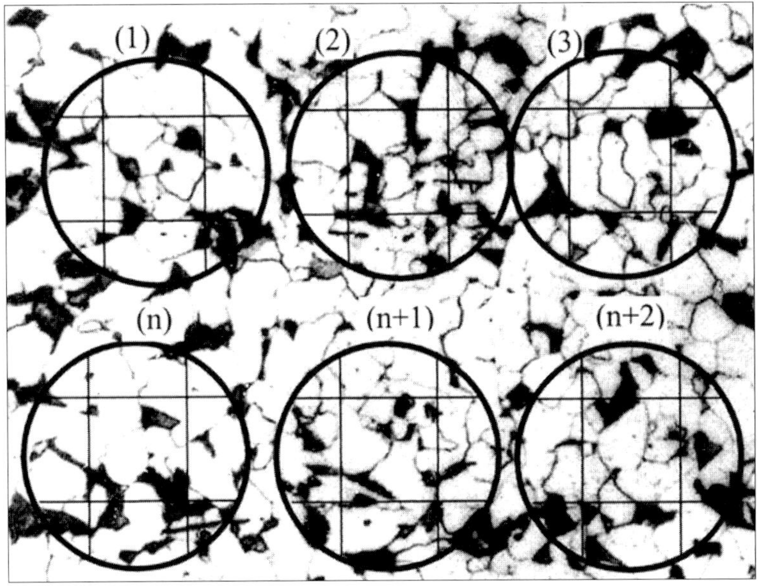

Fig. 3.2 Method of point counting using a 4 point graticule in the microscope eyepiece.

positions (n), $(n + 1)$, $(n + 2)$, it should be far enough from the first so that none of the counted areas overlap. As discussed in Chapter 1, when a point falls close to a boundary, rather than wasting time deciding whether it is just inside or just outside the boundary, it is counted as a half. Over a sufficient number of points, this should average out correctly.

Results

Using this method over a larger area of the microstructure shown in Fig. 3.1, the results in Table 3.1 were obtained when a 4-point graticule was moved to 100 different positions. In principle the number of points falling in pearlite (P_{pearl}) could range from 0 to 4, but in practice the numbers (n_i) with 3½ or 4 points were both zero.

The point fraction in each area ($P_{P(i)}$) in this example is clearly

$$P_{P(i)} = \frac{P_{(pearl)}}{4} \tag{3.4}$$

and

$$\overline{P}_P = \frac{\Sigma n_i\, P_{P(i)}}{\Sigma n_i} = \frac{17.50}{100} \tag{3.5}$$

$$= 0.1750$$

Table 3.1 Results of point counting using a 4-point graticule.

(1)	(2)	(3)	(4)	(5)
Number of Points in Pearlite (P_{Pearl})	Number of Areas Counted n_i	Point Fraction in Pearlite $P_{P(i)}$	$n_i P_{P(i)}$	$n_i (P_{P(i)})^2$
0	40	0	0	0
½	8	0.125	1.00	0.125
1	38	0.250	9.50	2.375
1½	4	0.375	1.50	0.5625
2	7	0.500	3.50	1.750
2½	2	0.625	1.25	0.78125
3	1	0.750	0.75	0.5625
3½	0	0.875	0	0
4	0	1.000	0	0
	Σn_i 100		$\Sigma n_i P_{P(i)}$ 17.50	$(\Sigma n_i P_{P(i)})^2$ 6.15625

Col 2 gives the raw data,
Col 3 is obtained from equation (3.4),
Col 4 and Col 5 are the products from Col 2 and Col 3.

The product $n_i(P_{P(i)})$ is shown in Column 4 of Table 3.1 and $n_i (P_{P(i)})^2$ is shown in Column 5, for calculation of the standard deviation from eqn (A.7) as

$$s^2 = \frac{6.15625 - 100(0.1750)^2}{99}$$

$$= 0.03125$$

$$s = 0.177$$

Hence, the standard error of $\bar{P}_P$ is

$$S(\bar{P}_P) = \frac{0.177}{\sqrt{100}} = 0.0177$$

and the 95% confidence limit is 0.0354.

Thus $\bar{P}_P = 0.175 \pm 0.035$

which may be rounded to $\bar{P}_P\ 0.18 \pm 0.04$

Comments

1. It is of interest to compare the measured value of $\bar{P}_P$ with the value expected in equilibrium for the steel composition.

For a steel of 1.35% Mn, it is expected that the equilibrium composition of pearlite will be ~ 0.72%C (Pickering (1983)). Also, with Ti mainly precipitated as nitrides/sulphides, carbon in solution may also be determined by the manganese content at ~0.01%C. The expected volume fraction of pearlite for the steel of 0.12%C can be calculated directly from the weight percentages, because the density of ferrite and pearlite are similar.

$$V_V \approx \frac{0.12-0.01}{0.72-0.01} \approx 0.155$$

This equilibrium value is less than the measured value, as expected for an air-cooled or normalised steel, but is still within the confidence limits of the experimental mean value.

2. The experimental value of the 95% confidence limit of ± 0.035 may be compared with the estimated value calculated from equation (2.3) as $2\ S(V_v)$. For $\bar{P}_p = 0.175$ and $P = 400$, this gives 95% confidence limits of ± 0.038. The two values are in good agreement.

3. It is noticeable from Table 3.1 that a large fraction of the areas (40 out of 100) had zero points falling in a pearlite colony. For lower volume fractions of ~0.1 it is therefore advantageous to use a 9-point graticule so that, on average, areas will have about one point falling in pearlite. For even lower volume fractions of ~0.05 a 16-point graticule may be used. However, for the present volume fraction of 0.175 these graticules are of no advantage. Even though fewer areas would need to be examined to obtain the same total number of points, having to count to more than 3 in any area is likely to lead to confusion and to end up taking a longer total measurement time for a given accuracy.

EXAMPLE 3.1.2 – USING MICROGRAPHS

Method

On a photomicrograph a complete grid of points is drawn, a small region of which is illustrated in Fig. 3.3. In the microscope, essentially the same effect is obtained with a point counting stage, which is electrically driven to move in steps each time a control key is pressed. Typically step size is 50 μm, but larger step sizes are available if coarse microstructures are to be counted. It is important that grid size, or the step size and spacing of traverses are sufficiently large for no two adjacent points to fall in the same pearlite colony. Again points are registered as 'in pearlite' (counts 1), 'in ferrite' (counts 0) and 'at boundary' (counts ½) to obtain the number of points in pearlite ($P_{(pearl)(i)}$) on each traverse. Examples are shown in Fig. 3.3. In practice longer traverses would be used for efficient counting.

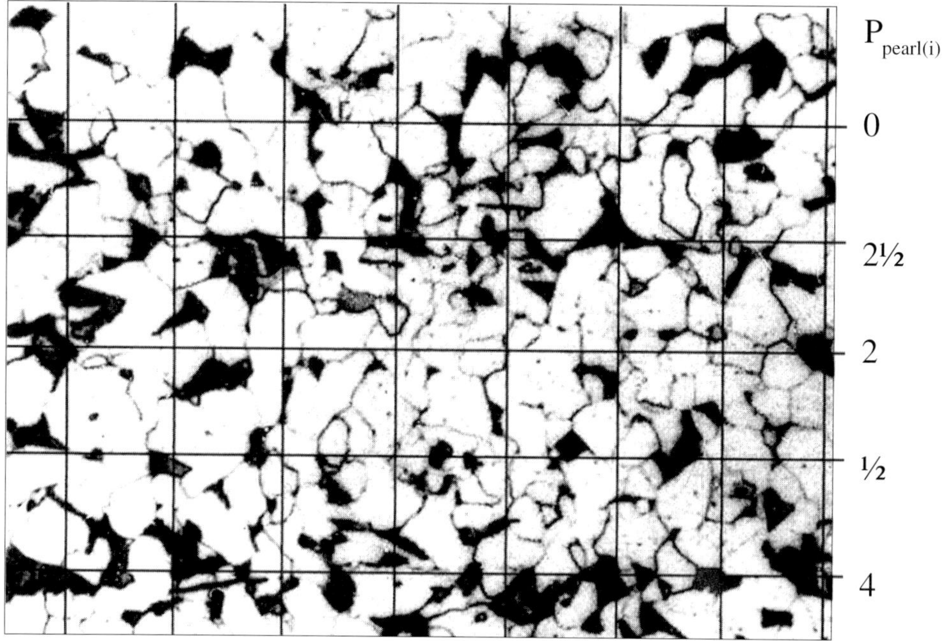

Fig. 3.3 Method of point counting using a grid of points.

Results

Using this method with 9 traverses, each of 46 points on a larger area of the microstructure shown in Fig. 3.1 gave the results in Table 3.2. (Actually 18 traverses of 23 points were measured but, for the example, results on two adjacent traverses have been summed).

The point fraction $P_{P(i)}$ on each traverse is simply obtained by dividing $P_{(pearl)(i)}$ by 46. Then

$$\bar{P}_P = \frac{\Sigma P_{(pearl)(i)}}{414} = \frac{\Sigma P_{P(i)}}{9} = 0.1763$$

Taking the numbers for each traverse as a measurement, calculation using equation (A.7) gives

$$s = 0.02355$$

Hence the standard error of $\bar{P}_P$ is

$$s\left(\bar{P}_P\right) = \frac{0.0236}{\sqrt{9}} = 0.00785$$

For 9 measurements (8 degrees of freedom), the value of $t_{95,\,8}$ is 2.306, leading to 95% confidence limits of ± 0.0181

Table 3.2 Results of point counting using points along traverses.

(1)	(2)	(3)
Traverse Number (i)	Number of Points in Pearlite $P_{(pearl)(i)}$	Point Fraction $P_{P(i)}$
1	8½	0.185
2	8½	0.185
3	7½	0.163
4	10	0.217
5	7½	0.163
6	7	0.152
7	7½	0.163
8	7	0.152
9	9½	0.207
	$\Sigma P_{(Pearl)}$ 73	$\Sigma P_{P(i)}$ 1.587

Col 2 gives raw data,
Col 3 is obtained as Col 2/46.

Thus $\bar{P}_P = 0.176 \pm 0.018$

which may be rounded to

$\bar{P}_P = 0.18 \pm 0.02$

Comments

1. The results of this method of measurement and Method 3.1 are in close agreement and are well within the confidence limits, as expected.
2. Other comments made about the results from Method 3.1 apply to these results, except that by chance the 95% confidence limits are significantly less than the value of 0.037 estimated from eqn (2.3).
3. If a Swift[†] point counting attachment is available on the microscope, it can be used to move the stage in predetermined steps of (say) 50 μm and enables point counting to be carried out rapidly on the microscope. Otherwise, the method of analysis is exactly the same as using traverses on micrographs, as in this example.

3.2 LINEAL ANALYSIS

Lineal analysis provides an alternative method to point counting for determining volume fraction. As discussed in Chapter 2, it becomes a more attractive alternative for use 'by hand' when the volume fraction of second phase is low. If the microstructure is nearly random, linear traverses may be carried out in any orientation to measure the length of

---

[†] Company Name

line occupied by the second phase. If several traverses are used, it is important to satisfy the 'golden rule' that the spacing between them is sufficiently large so that the same region of second phase is not intersected by adjacent traverses.

EXAMPLE 3.2.1 – USING MICROGRAPHS

Method

On a photomicrograph a series of traverse lines is drawn, as shown for a small region in Fig. 3.4. In order to measure the length of each traverse line that falls in pearlite, it is convenient to lay a sheet of paper on the micrograph, with the edge of the sheet coincident with the transverse line. The edges of the left-hand colony of pearlite are then marked as fine lines on the sheet of paper and the sheet is slid to the right until the right-hand mark coincides with the left-hand edge of the second colony from the left. The right edge of the colony is then marked as a fine line on the sheet of paper, and the sheet is again slid to the right etc. The process is repeated until the intercept sizes of all the colonies on the transverses have been accumulated. An example of the results of this procedure is shown in Fig. 3.5 for the first 3 traverses in Fig. 3.4. Measurement of the distance between the outer marks for each traverse gives directly the intercept lengths in pearlite ($L_{(pearl)(i)}$) shown in Fig. 3.5 and to the right of Fig. 3.4.

Using the above method for measuring the accumulated intercept lengths along a traverse also facilitates counting the number of individual colonies ($n(i)$) along the traverse. This is given by the number of spaces between the marks on the edge of the sheet of paper, as shown in Fig. 3.5 and repeated to the right of Fig. 3.4. These values are not necessary to determine the mean value of the line fraction, but they may be used to estimate the 95% confidence limits from eqn (2.9) and to determine the pearlite colony size. In practice, longer traverses would be used for efficient measurement.

Results

Using this method along 18 traverses each on 260 mm long on a larger micrograph printed at ×500 magnification gave the results in Table 3.3. For convenience in this table the results from two adjacent traverses have been summed to give effectively 9 traverses of 520 mm in length.

The length fraction in each traverse $L_L(i)$ shown in Column 4 is simply obtained by dividing the lengths in Column 2 by 520.
Then

$$\bar{L}_L = \frac{\sum L_L(i)}{9} = 0.183$$

Taking the numbers for each traverse as a measurement, calculation using equation (A.2) gives the standard deviation

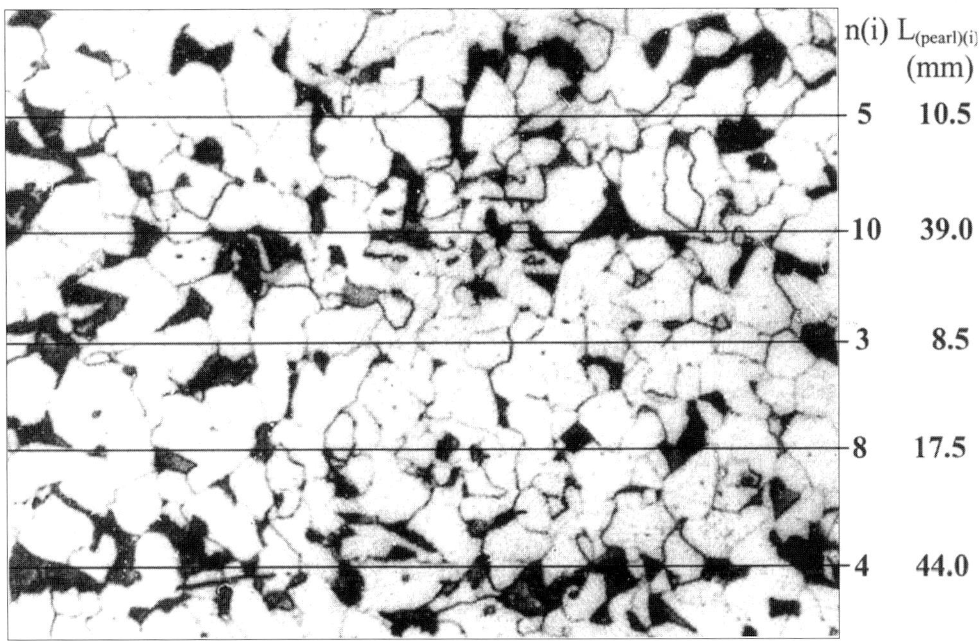

Fig. 3.4 Method of lineal analysis.

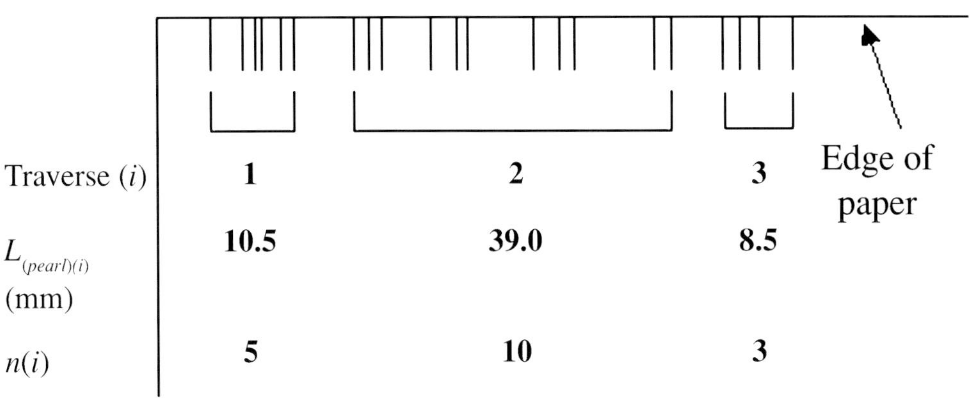

Fig. 3.5 Method of measuring $L_{(pearl)(i)}$ on traverses 1, 2 and 3 in Fig. 3.4.

Table 3.3 Results of lineal analysis.

(1)	(2)	(3)	(4)
Traverse Number (i)	Length in Pearlite $L_{(pearl)(i)}$ (mm)	Number of Pearlite Colonies $n(i)$	Length Fraction $L_L(i)$
1	104.0	23	0.200
2	87.0	45	0.167
3	97.5	13	0.188
4	121.5	36	0.234
5	110.0	54	0.212
6	105.5	27	0.302
7	82.0	41	0.158
8	78.0	32	0.150
9	70.5	22	0.136
	$\Sigma L_{(pearl)(i)}$ 826.0	$\Sigma n(i)$ 293	$\Sigma L_L(i)$ 1.648

Col. 2 and Col. 3 give the raw data,

Col. 4 is obtained as Col. 2/520.

$s = 0.0323$

Hence, the standard error of $\bar{L}_L$ is

$$S(\bar{L}_L) = \frac{0.0323}{\sqrt{9}} = 0.0108$$

For 8 degrees of freedom, the value of $t_{95,\,8}$ is 2.306, leading to 95% confidence limits of ±0.0248.

Thus

$$\bar{L}_L = 0.183 \pm 0.025$$

which may be rounded to

$$\bar{L}_L = 0.18 \pm 0.02$$

Comments

1. The mean value of $\bar{L}_L$ is in good agreement with the mean value of $\bar{P}_P$ determined from the same micrographs in Examples 3.1.1 and 3.1.2, indicating that with 95% confidence the true value of V_V lies in the range of 0.16 to 0.20.
2. The time taken to carry out measurements of 293 colonies of pearlite in Example 3.2.1 is longer than to count 400 points in Examples 3.1.1 and 3.1.2 for a similar accuracy of results, but at lower volume fractions the balance becomes more favourable for lineal analysis.

3. Measurements of the type illustrated in the example may by carried out more rapidly on the microscope, if a double micrometer microscope stage is available. This is a stage where one micrometer is used to drive a second stage mounted on it, which is also driven by a micrometer. One micrometer is then used to traverse the α-phase (pearlite in this example) and the other traverse the β-phase (ferrite in this example). At the end of each traverse, the reading on micrometer 1 gives $L_{(pearl)(i)}$ and the sum of the readings from both micrometers gives the total length of traverse, $L(i)$. The values of $L_L(i)$ are then analysed in the same way as in the example. If $n(i)$ is to be determined, the number of times micrometer 1 is adjusted for each traverse must be recorded.

4. When $n(i)$ is determined, the total number of measurements, $n = \Sigma n(i)$, can be used in eqn (2.9) to estimate the standard error of the mean. In the present example, from Column 3 in Table 3.3, $n = 293$. With $\bar{L}_L = 0.18$, substitution into eqn (2.9) gives

$$S(\bar{L}_L) = 0.18\left(\frac{2}{293}\right)^{1/2}(1 - 0.18)$$

$$= 0.012$$

Hence the estimated 95% confidence limits are $2 \times 0.012 = \pm0.024$, in close agreement with the value determined directly from the measurements.

5. When the value of the total number of colonies, n, is known as well as the total intercept length, $\Sigma L_{(pearl)(i)}$ in the present example, the mean pearlite colony size is simply obtained as

$$\bar{L}_P = \frac{826.0}{500} \times \frac{1}{293} = 5.6 \times 10^{-3}\,\text{mm}$$

This value is in excellent agreement with the value of $6.4 \pm 0.9 \times 10^{-3}$ mm determined by the alternative method of counting boundaries described in Example 4.3.1.

4. Size from Planar Sections

The grain size in single phase alloys, and grain size or 'colony' size in duplex structures such as ferrite-pearlite in carbon steels, α-β phases in brasses and recrystallised-unrecrystallised structures are usually of suitable scale to be determined by optical microscopy, or possibly scanning electron microscopy, from planar sections. Although it is obvious that grains and colonies are 3-dimensional features of metallurgical structures, it is conventional to report sizes measured on the planar sections, with no attempt to convert them to sizes of the 3-dimensional grains. The methods of measurement are based on counting numbers of boundaries per unit length or number of grains or colonies per unit area and are covered by ASTM standards (1999).

The commonly used method of estimating grain size for quality control purposes by comparison with standard charts is rapid, but is subjective and cannot be considered as a truly quantitative method, and so is not discussed further in this chapter.

4.1 LINEAR INTERCEPT GRAIN SIZE

The mean linear intercept, $\bar{L}$, is the average distance between grain boundaries along lines placed at random on the plane of polish. Lines of random orientation can be produced by superimposing circles on the microstructure as long as these are sufficiently large compared with the grain size. In practice, when grains are not equiaxed, it is preferable to measure along straight lines oriented in the three principal directions to obtain $\bar{L}_1$, $\bar{L}_2$ and $\bar{L}_3$, from which the mean linear intercept, $\bar{L}$, and additional microstructural parameters can be derived.

It is noteworthy that the methods of measurement involve counting the number of grain boundaries per unit length, $\bar{N}_L$, and defining $\bar{L}$ as

$$\bar{L} = \frac{1}{\bar{N}_L} \tag{4.1}$$

This is not the same mean value as would be found if the intercept lengths of individual grains were measured and averaged. Also on traverse lines of finite length it is important that the boundaries and not the grains are counted because the lines generally start in the middle of grains, giving one more grain than grain boundary along the traverse.

Fig. 4.1 shows the microstructure of a longitudinal section of a thick strip of Type 316 stainless steel that has been hot rolled and subsequently cold rolled by 28.5% reduction in thickness. As expected, the grains are elongated in the rolling direction, but the micrograph also shows other features that commonly occur in face-centred cubic metals

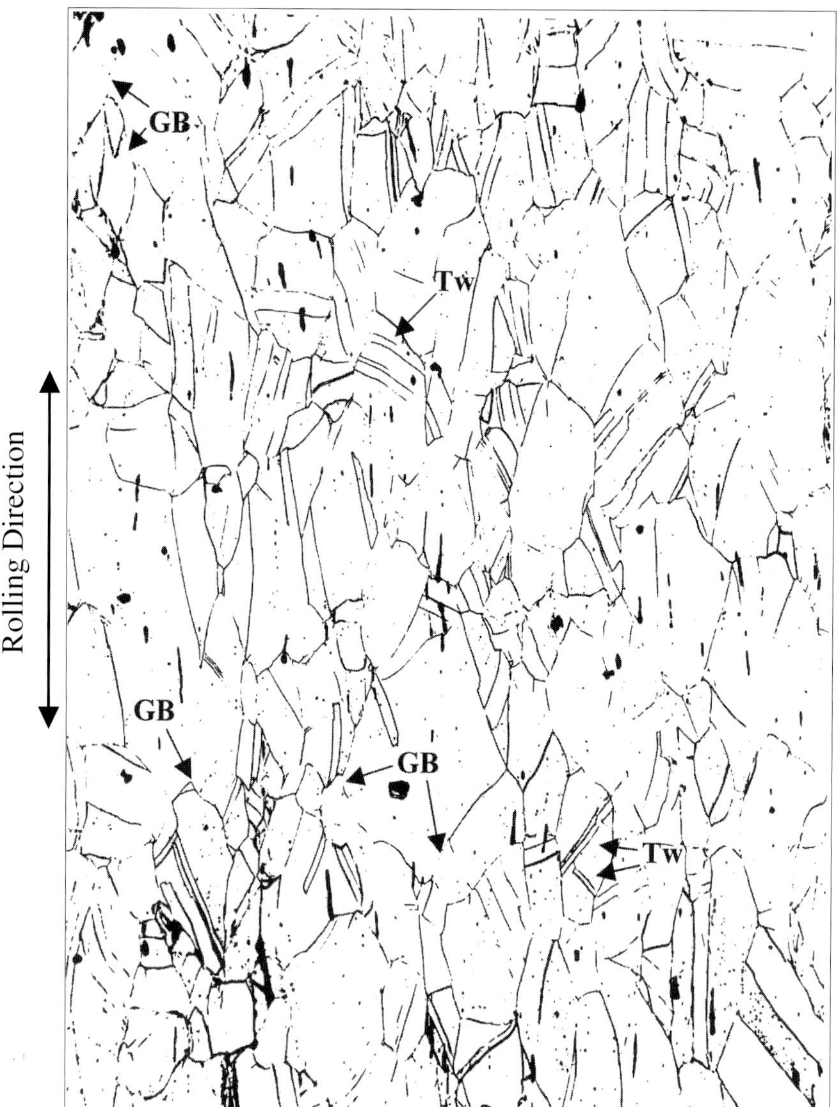

Fig. 4.1 Microstructure of Type 316 stainless steel cold rolled 28.5%, observed on a longitudinal section at a magnification of ×150.

and complicate the measurement of grain size. These are the annealing twins, examples of which are marked **Tw** in Fig. 4.1. These are present in most grains and have generally been bent by the cold deformation, although the twin boundaries in any grain are still parallel to each other. It is important that these features are recognised, because only grain boundaries and not twin boundaries must be counted to determine the grain size. A second complicating feature that frequently occurs with stainless steels, because of their resistance to etching, is that not all grain boundaries are etched up. The positions of these 'missing' boundaries can generally be recognised by the presence of cusps in adjacent boundaries that arise because of the surface tension effects of the boundaries. Several of these locations are indicated as **GB** in Fig. 4.1. It is important that these 'missing' boundaries are counted when determining the grain size. It is also apparent in the figure that twin boundaries do not generally produce cusps in the grain boundaries on which they end, because they have much lower surface energy than grain boundaries. This can also be a useful feature in distinguishing twin boundaries from grain boundaries.

EXAMPLE 4.1.1 – DIRECTIONAL LINEAR INTERCEPTS

Method

The linear intercepts of the deformed stainless steel are best measured along lines drawn on the print in the rolling direction (RD) and in the short transverse direction (STD), as illustrated in Figs 4.2 and 4.3. The boundaries that have been counted are marked with crosses on each traverse line. It can be seen that, because of the complications discussed earlier, there are some counted boundaries that may seem dubious grain boundaries. In fact, the situation is somewhat clearer on the original print and in practice it is frequently beneficial to measure directly on the screen of a projection microscope, where slight adjustment of the focus as counting is being carried out can help resolve ambiguity about the nature and presence of a boundary.

It should be noted in Figs 4.2 and 4.3 that the transverse lines have been spaced sufficiently far apart so that two adjacent lines only rarely cross the same grain. This is important for statistically significant sampling, but means that in order to obtain sufficient measurements, two other photographs of different areas of the sample had to be measured in addition to the area shown in Figs 4.2 and 4.3.

Results

Grain boundary counts on the photographs were taken along the full length of the transverse lines, i.e. 146 mm in the RD and 104 mm in the STD, at the print magnification of ×150. The numbers counted on each line are given in Table 4.1.

Note that the total lengths of traverse lines gives similar numbers of boundaries in each direction, to obtain similar relative errors.

If the conventional notation for defining the principal directions is used, i.e. 1 for the maximum principal strain (RD), 2 for the intermediate principal strain (the width direction, in which the strain is zero for plane strain rolling) and 3 for the minimum principal strain (STD) in which the strain is negative) then the linear intercept grain sizes $\overline{L}_1$ and $\overline{L}_3$ are obtained from the counts in the RD and STD as:

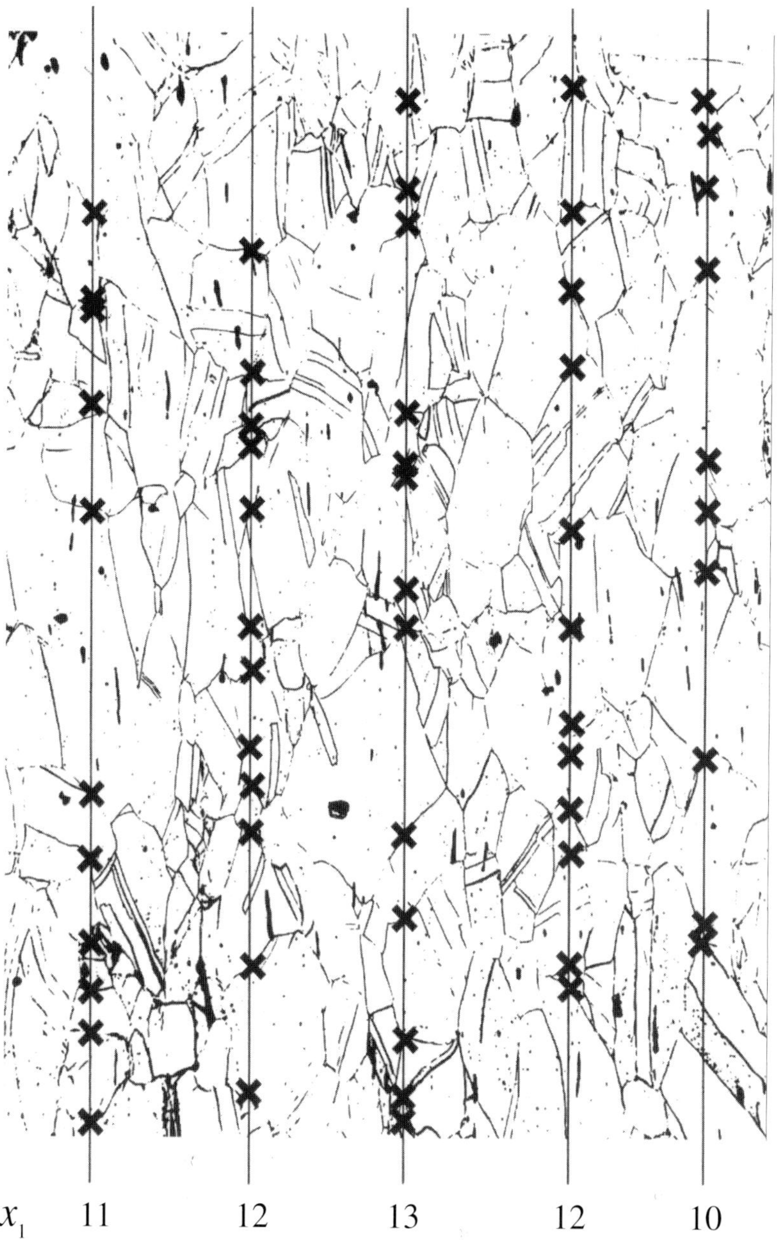

Fig. 4.2 Linear intercepts measured in the rolling direction of the cold rolled stainless steel, magnification ×150.

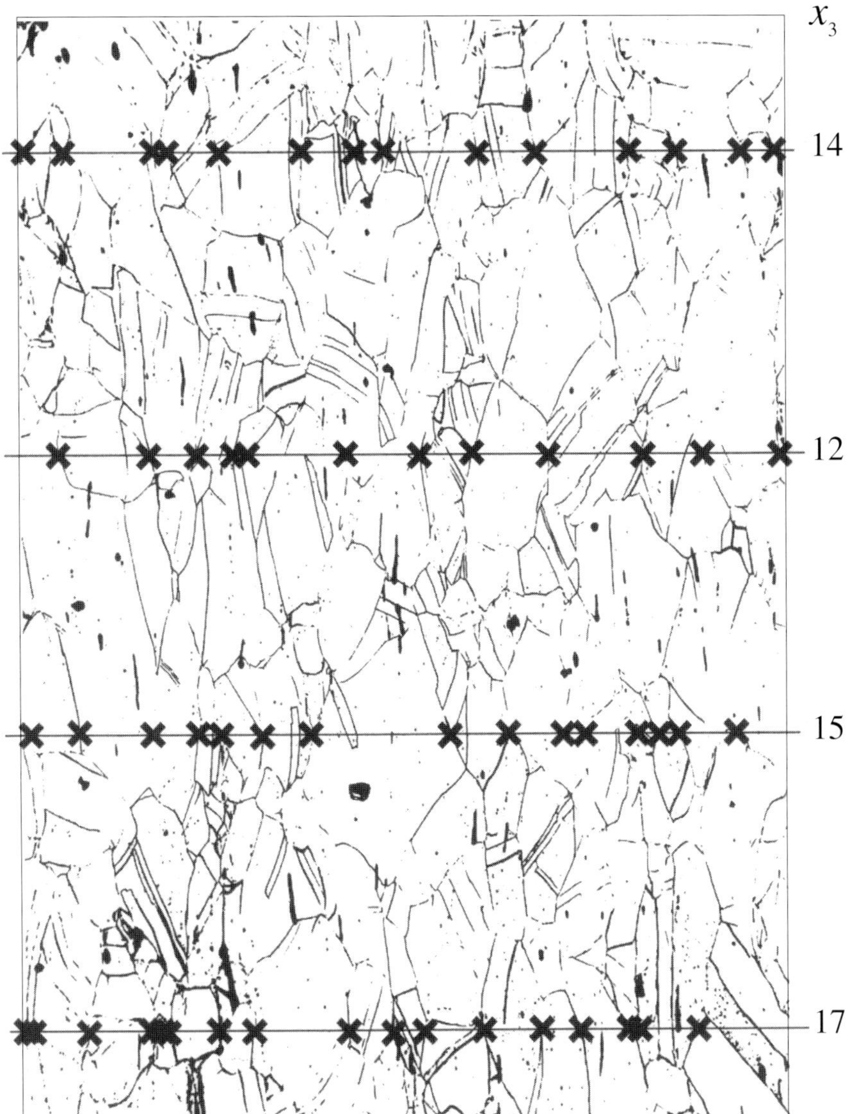

Fig. 4.3 Linear intercepts measured in the short transverse direction of the cold rolled stainless steel, magnification ×150.

Table 4.1 Results of the linear intercept measurements of the microstructure Figs 4.2 and 4.3 and two other fields in the sample.

	No. of Grain Boundaries in RD, x_1		No. of Grain Boundaries in STD, x_3
Figure 4.2	11	**Figure 4.3**	14
	12		12
	13		15
	12		17
	10		
Other	13	**Other**	18
Fields	14	**Fields**	16
	12		15
	11		17
	9		
	15		17
	12		12
	9		18
	12		17
	9		
Total	**174**		**188**

$$\bar{L} = \frac{\Sigma L}{M \times \Sigma x} \tag{4.2}$$

Where M is magnification and ΣL is the total length of the n traverses, i.e. 15 in the RD and 12 in the STD.

Hence, $\bar{L}_1 = \dfrac{146 \times 15}{150 \times 174} = 0.0839$ mm

and $\bar{L}_3 = \dfrac{104 \times 12}{150 \times 188} = 0.0443$ mm

The standard error can be estimated from eqn (2.10) in Chapter 2 as

$$\frac{S(\bar{L})}{\bar{L}} = \frac{0.65}{\sqrt{\Sigma x}} \tag{4.3}$$

giving

$$S(\bar{L}_1) = \frac{0.65}{\sqrt{174}} \times 0.0839 = 0.0041 \, \text{mm}$$

$$S(\overline{L}_3) = \frac{0.65}{\sqrt{188}} \times 0.0443 = 0.0021 \, \text{mm}$$

Because of the large number of measurements, the value of $t_{95,\,n-1} \simeq 2$ and the 95% confidence limits of $\overline{L}_1$ and $\overline{L}_3$ are ~0.0082 and 0.0042 mm respectively. The mean values should therefore be rounded to the number of significant figures for reporting the grain sizes i.e.

$$\overline{L}_1 = 0.084 \pm 0.008 \, \text{mm}$$

and $\quad \overline{L}_3 = 0.044 \pm 0.004 \, \text{mm}$

Alternative Procedure

The counts on each of the n traverse lines can be considered as separate samples, and the results are then analysed using the standard statistical methods summarised in the appendix.

From eqn (A.1)

$$\overline{x}_1 = \frac{\Sigma x_1}{n_1} = \frac{174}{15} = 11.60$$

$$\overline{x}_3 = \frac{\Sigma x_3}{n_3} = \frac{188}{12} = 15.67$$

and from eqn (A.2), using a calculator or spreadsheet, the standard deviations are

$$s\,(x_1) = 1.805$$

$$s\,(x_3) = 2.103$$

Hence from eqn (A.8) the standard errors of the means are

$$S(\overline{x}_1) = \frac{1.805}{\sqrt{15}} = 0.466$$

$$S(\overline{x}_3) = \frac{2.103}{\sqrt{12}} = 0.607$$

These values are converted to linear intercepts as

$$\overline{L} = \frac{L}{M} \cdot \frac{1}{\overline{x}} \tag{4.4}$$

giving $\bar{L}_1 = \dfrac{146}{150} \cdot \dfrac{1}{11.60} = 0.0839\,\text{mm}$

$\bar{L}_3 = \dfrac{104}{150} \cdot \dfrac{1}{15.67} = 0.0442\,\text{mm}$

Also,

$$\frac{S(\bar{L})}{\bar{L}} = \frac{S(\bar{x})}{\bar{x}} \tag{4.5}$$

giving

$$S(\bar{L}_1) = \frac{0.466}{11.6} \times 0.0839 = 0.0034\,\text{mm}$$

$$S(\bar{L}_3) = \frac{0.607}{15.67} \times 0.0442 = 0.0017\,\text{mm}$$

From Fig. A4/Table A1, with 15 and 12 measurements (14 and 11 degrees of freedom) the values of $t_{95,\,n\text{-}1}$ are 2.145 and 2.201, leading to 95% confidence limits of $\bar{L}_1$ and $\bar{L}_3$ of 0.0073 and 0.0037, respectively.

Again, rounding the mean values to the number of significant figures gives

$$\bar{L}_1 = 0.084 \pm 0.008\,\text{mm}$$
$$\bar{L}_3 = 0.044 \pm 0.004\,\text{mm}$$

Comments

1. As expected, the mean values obtained by the alternative methods of analysis are identical. Although there are small differences in the confidence limits, because the latter values are derived from the actual measurements whereas the former are generic estimates, the differences are not significant. While the alternative method of analysis is preferable in principle, the estimated confidence limits are clearly satisfactory in practice.

2. It is important that grain sizes are rounded to the significant figures for reporting, but as discussed in later examples, the complete values should be used in any further computations of grain size and shape parameters, with rounding to the significant figures of the parameters carried out as the final step in the analysis.

3. Linear intercept grain sizes are commonly reported in micrometers as

$$d_1 = 84 \pm 8\;\mu\text{m} \text{ and}$$

$$d_3 = 44 \pm 4\;\mu\text{m}$$

4.   Frequently an overall mean linear intercept grain size and the aspect ratio are of interest to relate to the microstructure before deformation. These are considered in the next examples.

EXAMPLE 4.1.2 – OVERALL MEAN LINEAR INTERCEPT

The volume of a grain is unchanged by plastic deformation so the overall mean linear intercept for the structure is defined as

$$\overline{L} = (\overline{L}_1 \overline{L}_2 \overline{L}_3)^{\frac{1}{3}} \tag{4.6}$$

In general, this requires measurements to be made on two planes, but for plane strain deformation $\overline{L}_2$ remains constant and is equal to $\overline{L}$, hence

$$\overline{L} = (\overline{L}_1 \overline{L}_3)^{\frac{1}{2}} \tag{4.7}$$

Results

Equation (4.7) is appropriate for the cold rolled stainless steel, so from the results of Example 4.1.1.

$$\overline{L} = (0.0839 \times 0.0442)^{\frac{1}{2}}$$

$$= 0.0609 \text{ mm}$$

The 95% confidence interval for $\overline{L}$ can be found by applying the method of propagation of errors described in the Appendix. In this case eqn (A.12) becomes

$$\overline{L} = \overline{L}_1^{\frac{1}{2}} \overline{L}_3^{\frac{1}{2}} \tag{4.8}$$

and eqn (A.13) becomes

$$(S(\overline{L}))^2 = (S(\overline{L}_1))^2 (\frac{1}{2} \cdot \overline{L}_1^{-\frac{1}{2}} \overline{L}_3^{\frac{1}{2}})^2 + (S(\overline{L}_3))^2 (\frac{1}{2} \cdot \overline{L}_1^{\frac{1}{2}} \overline{L}_3^{-\frac{1}{2}}) \tag{4.9}$$

$$= (S(\overline{L}_1))^2 \frac{1}{4} \cdot \frac{\overline{L}_3}{\overline{L}_1} + (S(\overline{L}_3))^2 \frac{1}{4} \cdot \frac{\overline{L}_1}{\overline{L}_3} \tag{4.10}$$

substituting the earlier values gives

$$(S(\bar{L}))^2 = (0.0034)^2 \frac{1}{4} \cdot \frac{0.0442}{0.0839} + (0.0017)^2 \frac{1}{4} \cdot \frac{0.0839}{0.0442}$$

$$= 1.523 \times 10^{-6} + 1.371 \times 10^{-6}$$

$$= 2.894 \times 10^{-6}$$

$$S(\bar{L}) = 1.7 \times 10^{-3}$$

giving a 95% confidence limit of about 0.004 mm, so that rounding to significant figures gives

$$\bar{L} = 0.061 \pm 0.004 \, \text{mm}$$

Alternative Method

Alternatively, grain counting can be carried out around circles to obtain $\bar{L}$ directly. The method is illustrated in Fig. 4.4 with the points of intersection of grain boundaries with the 150 mm circumference circles indicated by crosses. The method is subject to the same ambiguities as in Example 4.1.1.

Results

For the 4 circles in Fig. 4.4 the numbers of intersections are 18, 18, 18, 19. As for Example, 4.1.1, additional measurements were made on two other areas to give a total of 12 circles of length

$$\Sigma L = \frac{12 \times 150}{150} \, \text{mm}$$

on the specimen, along which a total of 211 grain boundaries were counted. Following the same analysis procedure as for Example 4.1.1, eqn (4.2) becomes

$$\bar{L} = \frac{150 \times 12}{150 \times 211} = 0.0569 \, \text{mm}$$

and from eqn (2.10) the 95% confidence limit is 0.005 mm, giving

$$\bar{L} = 0.057 \pm 0.005 \, \text{mm}$$

Comments

1. It can be seen that the results of the two methods of measuring $\bar{L}$ give results in close agreement within the combined confidence limits. However, the method of counting round circles cannot provide any information about grain shape.

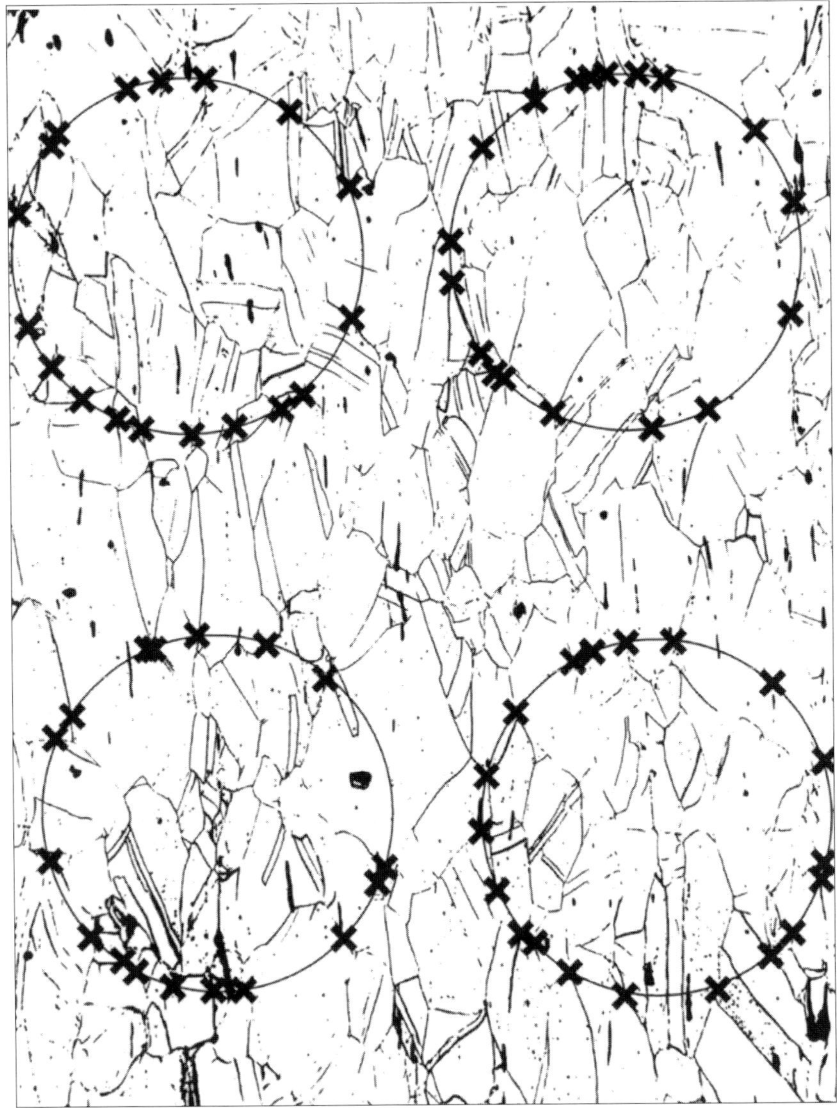

Fig. 4.4 Linear intercepts measured with circles (150 mm circumference) on a longitudinal section of the cold rolled stainless steel, magnification ×150.

2. In ASTM Standard E112 a pattern of test lines comprising concentric circles of circumference 250.0, 166.7 and 83.3 mm is permitted. For equiaxed structures, an appropriate magnification can be found to satisfy the condition that adjacent test lines do not intersect the same grain, but this is more difficult for elongated structures, so separate circles have been used in the example.

EXAMPLE 4.1.3 – GRAIN ASPECT RATIO

Results

The mean aspect ratio of the grains in the section can be obtained simply as the ratio of linear intercepts, so from the results of Example 4.1.1 the aspect ratio

$$\bar{r} = \frac{\bar{L}_1}{\bar{L}_3} = \frac{0.0839}{0.0442} = 1.898 \approx 1.9$$

In the case of $\bar{r}$, the confidence limits cannot be calculated from the confidence limits of $\bar{L}_1$ and $\bar{L}_3$, because eqn A.11 for the propagation of errors assumes that the variables are independent. Inspection of Fig. 4.1 shows clearly that for individual grains L_1 and L_3 are closely correlated i.e. grains are generally small or large in both directions.

Alternative Method

If the value of $\bar{r}$ and its confidence limits are required, the tangent diameters of individual grains must be measured, as illustrated in Fig. 4.5. Tangent diameters are defined as the maximum distance between parallel lines that just contact the grains.

To measure $\bar{r}$, a number of grains should be selected at random. In Fig. 4.5 ten grains have been selected, for which there is no ambiguity in the position of the boundaries. For these grains the tangent diameters in the 1 and 3 directions have been measured to the nearest 0.5 mm, and are reported in Table 4.2.

Table 4.2 Tangent diameters of the grains illustrated in Fig. 4.5.

(1)	(2)	(3)	(4)
Grain	L_1 (mm)	L_3 (mm)	r
1	24.5	11.0	2.23
2	24.0	13.0	1.85
3	21.0	9.0	2.33
4	14.0	6.0	2.33
5	22.5	12.5	1.80
6	30.0	17.5	1.71
7	12.0	6.5	1.85
8	23.0	10.5	2.19
9	18.0	11.5	1.57
10	18.0	14.5	1.24

Col 2 and Col 3 are raw data,
Col 4 is obtained as Col 2 ÷ Col 3.

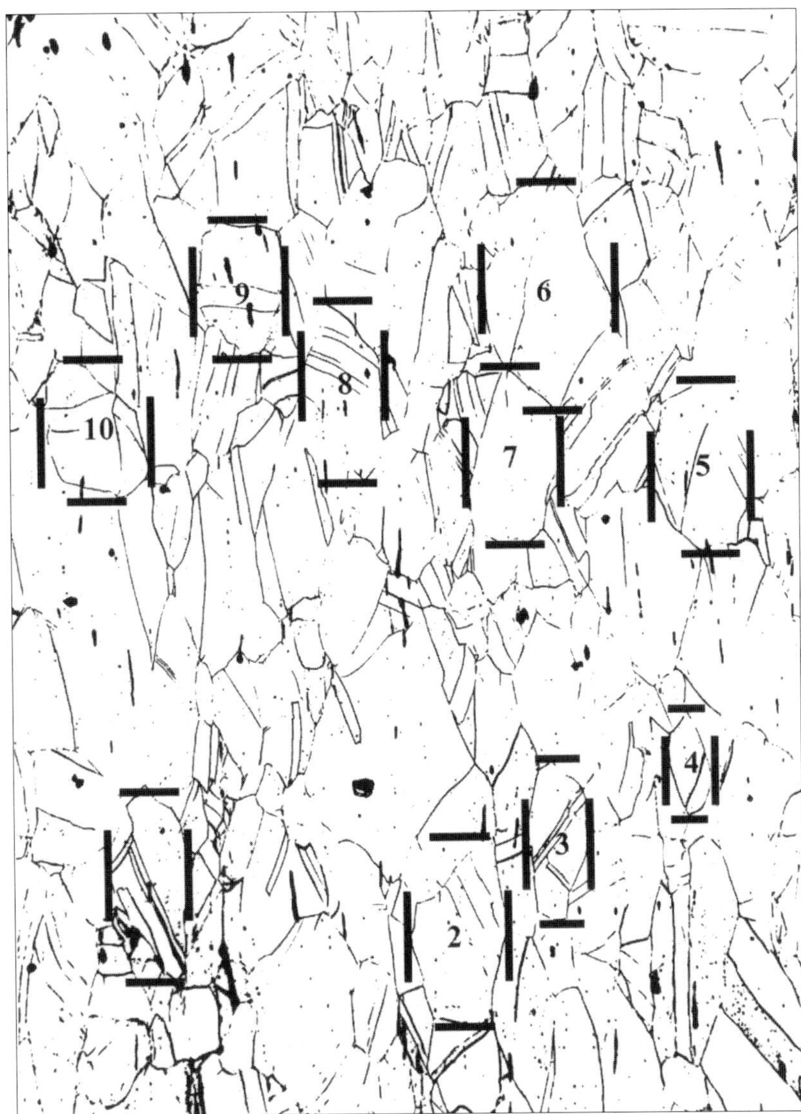

Fig. 4.5 Aspect ratio using tangent diameter measurements of ten grains of the cold rolled stainless steel, magnification ×150.

Results

Analysis of the data in Table 4.2 using the standard statistical procedures summarised in the Appendix leads to

$$\bar{r} \quad\quad = 1.910$$
$$s(r) \quad = 0.359$$
$$S(\bar{r}) \quad = 0.113$$
$$95\%CL = 0.26$$

Thus $\bar{r} = 1.91 \pm 0.26$
Measurement of a further 20 grains leads to
$\bar{r} = 1.92 \pm 0.12$

Comments

1. The two methods clearly give consistent values of $\bar{r}$.
2. The value of confidence limit of ±0.12 obtained from measurement of tangent diameters of only 30 grains compares with a calculated value of ±0.3 if the confidence limits from 174 grains for $\bar{L}_1$ and 188 grains for $\bar{L}_3$ are incorrectly used to calculate confidence limits by propagation of errors. This clearly illustrates the importance of the correlation between L_1 and L_3 for each grain.
3. For plane strain rolling, this grain aspect ratio should directly reflect the reduction of the work piece. Assuming a cube of side, l, before rolling, the length of the sides after a fractional reduction, R, becomes

$$l_1 = l/(1-R)$$
$$l_2 = l$$
$$l_3 = l/(1-R)$$

the aspect ratio is therefore

$$r = \frac{l_1}{l_3} = \frac{1}{(1-R)^2} \tag{4.11}$$

In the present example the cold rolling reduction is 28.5%, i.e. $R = 0.285$ and

$$r = \frac{1}{(0.715)^2} = 1.96$$

This is close to the value from the grain shape measurements, indicating that the initial grain size $\bar{L}$ was nearly equiaxed, as expected after hot rolling at a high enough temperature for full recrystallisation to have taken place.

EXAMPLE 4.1.4 – SURFACE AREA PER UNIT VOLUME

One of the advantages of measuring mean linear intercept grain size is that the number of intercepts per unit length, N_L, is directly related to the grain boundary area per unit volume, S_V, which is an important parameter in determining the kinetics of many transformations because grain boundaries provide preferential sites for nucleation.

When the grains are equiaxed,

$$S_V = 2N_L = 2/\overline{L} \tag{4.12}$$

After deformation, S_V is a weighted average of the values of $\overline{N}_L$ in each of the three principal directions (Underwood (1970)). Using the conventional notation described earlier in Example 4.1.1, with the maximum principal strain in the 1 direction and the minimum principal strain in the 3 directions, the relationship may be generalised as

$$S_V = 0.429\overline{N}_{L1} + 0.571\overline{N}_{L2} + \overline{N}_{L3} \tag{4.13}$$

Eqn. (4.13) reverts to eqn (4.12) when $\overline{N}_{L1} = \overline{N}_{L2} = \overline{N}_{L3}$.

Results

For the stainless steel in Example 4.1.2, applying eqn (4.12) to the equiaxed grain size of 0.061 mm before deformation gives

$$S_V = \frac{2}{0.061}$$

$$= 32.8 \text{ mm}^2 \text{ mm}^{-3}$$

After cold rolling, the measurements on the stainless steel in Example 4.1.1 give

$$L_1 = \quad 0.084 \text{ mm}$$
$$\overline{L}_2 = \overline{L} = 0.061 \text{ mm}$$
$$\overline{L}_3 = \quad 0.044 \text{ mm}$$

Substituting these values into eqn (4.13) gives

$$S_V = \frac{0.429}{0.084} + \frac{0.571}{0.061} + \frac{1}{0.044}$$
$$= 37.2 \text{ mm}^2 \text{ mm}^{-3}$$

Comments

1. Deformation increases the grain boundary area per unit volume, and this effect becomes more and more important as deformation increases. For the above example the increase in S_V by a factor of 1.13 by rolling is equivalent to reducing the equiaxed

grain size from 0.061 to 0.054 mm. In general, additional effects on transformation kinetics arise from the dislocation structure in a worked metal.

2.   If it is imagined that the microstructure in Fig. 4.1 arises from axisymmetric cold drawing instead of plane strain rolling, then

$$\overline{L}_1 = 0.084 \, \text{mm}$$

$$\overline{L}_2 = \overline{L}_3 = 0.044 \, \text{mm}$$

and the original grain size would have been

$$\overline{L} = (0.084)^{\frac{1}{3}} (0.044)^{\frac{2}{3}} = 0.055 \, \text{mm}$$

giving

$$S_{VO} = 2/0.055 = 36.6 \, \text{mm}^2 \, \text{mm}^{-3}$$

and

$$S_V = \frac{0.429}{0.084} + \frac{1.571}{0.044}$$
$$= 40.82 \, \text{mm}^2 \, \text{mm}^{-3}$$

This is an increase by a factor of 1.12.

If, conversely, it is imagined that the microstructure in Fig. 4.1 arises from axisymmetric upset forging, then

$$\overline{L}_1 = \overline{L}_2 = 0.084 \, \text{mm}$$

$$\overline{L}_3 = 0.044 \, \text{mm}$$

and the original grain size would have been

$$\overline{L} = (0.084)^{\frac{2}{3}} (0.044)^{\frac{1}{3}} = 0.068 \, \text{mm}$$

giving

$$S_{VO} = 2/0.068 = 29.5 \, \text{mm}^2 \, \text{mm}^{-3}$$

and

$$S_V = \frac{1}{0.084} + \frac{1}{0.044}$$
$$= 34.6 \, \text{mm}^2 \, \text{mm}^{-3}$$

This is an increase by a factor of 1.17.

Thus the appearance of elongated grains of a given aspect ratio in one section gives very different surface area per unit volume, depending on the mode of deformation that produced the elongation. In general, deformed microstructures should therefore be measured on two planes to obtain true values for $\overline{L}_1$, $\overline{L}_2$ and $\overline{L}_3$.

4.2 ASTM GRAIN SIZE

The ASTM grain size number, g, was originally defined from the number of grains per square inch at a magnification of ×100, but is now related to the number of grains per millimetre squared ($\bar{N}_A$) measured at any appropriate magnification as

$$g = -2.954 + 3.322 \log \bar{N}_A \tag{4.14}$$

EXAMPLE 4.2.1 – ASTM GRAIN SIZE

Method

To determine the value of $\bar{N}_A$ the numbers of grains in appropriate areas must be counted. The method is illustrated in Fig. 4.6, which is an enlargement of one of the circles in Fig. 4.4. Because it is important that all grains in the area are counted only once, the microstructure in Fig. 4.6(a) has been traced to remove ambiguities and to ensure that there is a complete network of boundaries, as shown in Fig. 4.6(b).

The grains that are fully enclosed within the circle are marked with a '1' and those that intersect the circle are marked with a '•'. All the grains marked '•' are counted as 1/2, independent of whether a major part or only a minor part of an individual grain is inside the circle, because on average the intersected grains will be half in the circle. In the example in Fig. 4.6 there are 8 grains marked '1' and 16 grains marked '•'.

Results

The circle is 150 mm circumference at the magnification of × 150 in Fig. 4.4, so the radius

$$r = \frac{150}{2\pi} \cdot \frac{1}{150} = 0.159 \, \text{mm}$$

and the true area is

$$A = \pi \times 0.159^2 = 0.080 \, \text{mm}^2$$

Note, the area of the circle on the micrograph is

$$A \times M^2 = 1800 \, \text{mm}^2$$

The grain counts give a value of

$$N_{A(i)} = \frac{(8 + 16/2)}{0.080} = 200 \, \text{mm}^{-2}$$

Counts were made in all four circles in Fig. 4.4 and in eight other circles to give a total of 194 grains. Thus,

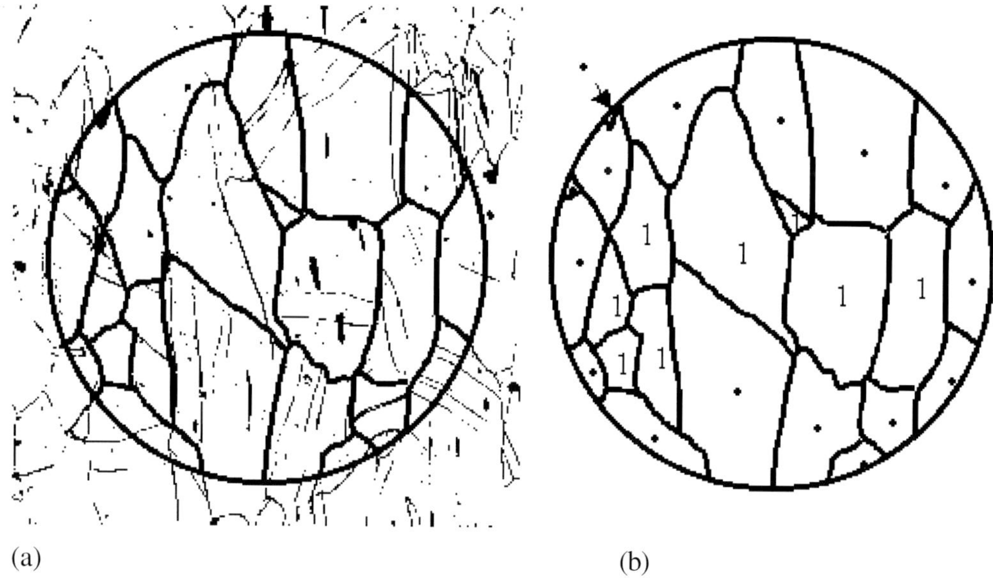

Fig. 4.6 (a) An enlargement of the top left circle in Fig. 4.4 with the position of grain boundaries superimposed on the micrograph. (b) the position of the grain boundaries in (a) showing the marks used in counting the grains.

$$\overline{N}_A = \frac{194}{12 \times 0.080} = 202.1 \, \text{mm}^{-2}$$

From eqn (2.11), the expected relative standard error

$$\frac{S(\overline{N}_A)}{\overline{N}_A} = \frac{1.03}{\sqrt{194}} = 0.074$$

so the 95% confidence limits of $\overline{N}_A$ are $\pm\, 2 \times 0.074 \times 202.1 = \pm\, 29.9 \, \text{mm}^{-2}$

Alternatively, counting each of the 12 circles as a measurement, and following the same statistical procedure as for Example 4.1.1, leads to a value of standard deviation

$$s(N_A) = 45.9 \, \text{mm}^{-2}$$

Hence, for $n = 12$ the standard error is

$$S(\overline{N}_A) = s(N_A)/\sqrt{n} = 13.21 \, \text{mm}^{-2}$$

With 11 degrees of freedom, $t_{95,\,11} = 2.201$ and the 95% confidence limits of $\overline{N}_A$ are $\pm 29.1 \, \text{mm}^{-2}$.

Again these values of confidence limit are in close agreement with the value of $\pm\,29.9$ mm^{-2} from eqn (2.11), and in both cases rounding the result to significant figures gives

$$\bar{N}_A = 202 \pm 30 \text{ mm}^{-2}$$

Substituting these values into eqn (4.14) gives

$$\bar{g} = 4.70$$

with the limits 4.90 and 4.47, which is approximately

$$\bar{g} = 4.7 \pm 0.2$$

From eqn (2.12), it is expected that

$$S(\bar{g}) = \frac{1.49}{194} = 0.107$$

giving a value of for the confidence limits of $\bar{g}$, which is again in close agreement with the value of ± 0.2 calculated from the confidence limits of $\bar{N}_A$.

Comments

1. If, as illustrated here, the grain boundaries have to be traced to ensure that the complete network of grains is counted, it will take much more effort to determine $\bar{N}_A$ than to determine the number of grains per unit length in Examples 4.1.1 and 4.1.2. Also, it is clear that when numbers of grains per unit area are counted, no information is obtained about the grain aspect ratio. However, if tracings of the sort shown in Fig. 4.6(b) have to be made, these provide ideal images for automatic quantitative analysis, which provides additional information about the microstructures.

2. An alternative measure of grain size that is sometimes used is

$$\sqrt{\bar{A}} = 1/\sqrt{\bar{N}_A} \tag{4.15}$$

    For the present example, this leads to a grain size of

$$\sqrt{\bar{A}} = \frac{1}{\sqrt{202}} = 0.070 \text{ mm}$$

    This value, which can be considered as the length of the side of a square grain, would have little physical meaning. However, if the idealised grain structure in the volume is considered to be uniform sized tetrakaidecahedra (TKDs), it can be shown that:

$$\sqrt{\bar{A}} = 1.15\bar{L} \tag{4.16}$$

    Hence the estimated value of the mean linear intercept grain size is

$$\bar{l} = 0.061\,\text{mm}$$

In practice, for real microstructures, Pereira de Silva (1966) found that

$$\sqrt{\bar{A}} = 1.07\bar{L} \qquad\qquad (4.17)$$

This is in reasonable agreement with eqn (4.16), and would lead to a value of

$$\bar{l} = 0.065\ mm$$

Both values are within the 95% confidence limits of the overall mean linear intercept grain size determined in Example 4.1.2, with the one for the idealised structure in the volume actually being closer.

3. The idealised structure of uniform TKDs in the volume enables standard conversions to be made between ASTM grain size number, number of grains per unit area (mm^{-2}) and mean linear intercept grain size (mm) (ASTM standard E112). These are based on the relationship for equivalent ASTM number

$$\bar{g}e = -3.356 - 6.644\log_{10}\bar{L} \qquad\qquad (4.18)$$

Thus, when measurements are made by hand it is easier and quicker to measure $\bar{L}$ in mm and convert to ASTM number. In the present example,

$$\bar{g}e = -3.356 - 6.644\log_{10}(0.061 \pm 0.004)$$
$$= 4.7 \pm 0.2$$

The value is coincidentally in precise agreement with the value measured directly by determining $\bar{N}_A$.

4.3 COLONY SIZE OR GRAIN SIZE IN DUPLEX STRUCTURES

Duplex structures of two phases frequently occur in metallurgical microstructures. One classical example is the ferrite/pearlite structure of air cooled low carbon steels, in which the pearlite colony size and the ferrite grain size are both of importance in determining the mechanical properties.

EXAMPLE 4.3.1 – COLONY AND GRAIN SIZE IN A FERRITE-PEARLITE STRUCTURE

Method

Fig. 3.1 shows an example of the ferrite-pearlite structure in a microalloyed steel. This was used in Example 3.1.1 and 3.1.2 to determine the volume fraction of pearlite as 0.18 ± 0.03. The same micrograph is used in this example to count the number of boundaries

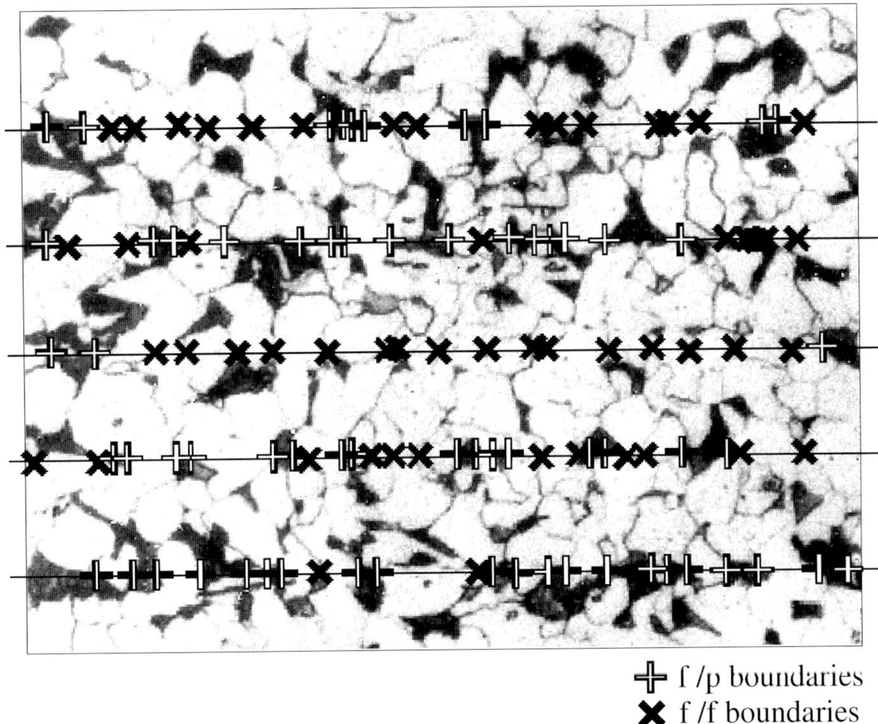

Fig. 4.7 Method of counting the f/p and f/f grain boundaries in microalloyed steel.

per unit length along the same traverse lines as used for point counting in Fig. 3.3. The method requires separate counts of the number of ferrite/pearlite (f/p) boundaries, $n_{p(i)}$, and the number of ferrite/ferrite (f/f) grain boundaries, $n_{\alpha(i)}$, to be made along each of the traverses, as illustrated in Fig. 4.7.

Results

Using the same 9 traverses as for point counting in Example 3.1.2 leads to the results for the numbers of ferrite/pearlite and the numbers of ferrite/ferrite boundaries shown in columns 4 and 6 of Table 4.3 respectively. For convenience the point counting results in Example 3.1.2 are repeated in Columns 2 and 3 of Table 4.3.

Each pearlite colony has two boundaries, so the number of colonies on each traverse is half the number of boundaries. Also the line fraction occupied by pearlite can be estimated from the point fraction on each traverse, $P_{p(i)}$. Hence, the number of pearlite colonies per unit length on each traverse is

Table 4.3 Results of point counting and counting boundaries.

(1)	(2)	(3)	(4)	(5)	(6)	(7)
Traverse Number (i)	No. of Points in Pearlite $P_{(pearl)(i)}$	Point Fraction $P_{P(i)}$	No. of Pearlite Boundaries $n_{P(i)}$	No. Per Length $Np(i)$ mm^{-1}	No. of Ferrite Boundaries $n_{\alpha(i)}$	No per Length $N_{\alpha(i)}$ mm^{-1}
1	8½	0.185	41	125.9	37	80.2
2	8½	0.185	48	147.4	30	75.3
3	7½	0.163	58	202.2	35	86.9
4	10	0.217	54	141.4	35	90.0
5	7½	0.163	55	191.7	36	86.2
6	7	0.152	46	171.9	37	80.4
7	7½	0.163	48	167.3	37	82.8
8	7	0.152	54	201.9	36	84.4
9	9½	0.207	50	137.2	31	80.2
Total	**73**	**1.587**	**454**	**1486.9**	**314**	**746.4**

Col. 2 Raw data from Example 3.1.2,
Col. 3 Results of point fractions from Example 3.1.2,
Col. 4 Experimental results for number of ferrite/pearlite boundaries,
Col. 5 Number of pearlite colonies per unit length obtained from Col. 3, Col 4 and eqn (4.19),
Col. 6 Experimental results for the number of ferrite/ferrite boundaries,
Col. 7 Number of ferrite grains per unit length obtained from Col. 3, Col. 4, Col. 6 and eqn (4.20).

$$N_{p(i)} = \frac{n_{p(i)}}{2P_{p(i)}.L} \tag{4.19}$$

where L is the length of each traverse. In the present example, the traverses on the micrographs are 440 mm and the magnification is ×500, so L = 0.88 mm = 880 µm.

Substitution of the values of $P_{p(i)}$ from Column 3 and of $n_{p(i)}$ from Column 4 of Table 4.3 into eqn (4.19) leads to the values of $N_{p(i)}$ given in Column 5.

Taking the mean value of $\bar{P}_p$, calculated from the total in Column 3, as discussed in Exercise 3.1.2, gives 0.176. Thus from the total in Column 4, by substitution into eqn (4.19).

$$\bar{N}_p = \frac{454}{2 \times 0.176 \times 9 \times 0.88}$$

$$= 162.9 \, \text{mm}^{-1}$$

Thus, the mean pearlite colony size

$$\bar{L}_p = \frac{1}{\bar{N}_p} = 0.00614 \, \text{mm}$$

$$= 6.14 \, \mu\text{m}$$

Alternatively, the number of traverses can simply be divided by the total in column 5 to give

$$\bar{L}_p = 9/1486.9 = 6.05 \ \mu m$$

The small difference in $\bar{L}_p$ arises from the difference in weighting of each traverse in the two methods.

In order to obtain the confidence limit of the mean, each traverse is considered as a measurement and from eqn (A.2), the standard deviation of $N_{p(i)}$.

$$s = 28.9 \ mm^{-1}$$

Hence the standard error of $\bar{N}_p$ is

$$S(\bar{N}_p) = \frac{28.9}{\sqrt{9}} = 9.63 \ mm^{-1}$$

The relative standard errors

$$\frac{S(\bar{L}_p)}{\bar{L}_p} = \frac{S(\bar{N}_p)}{\bar{N}_p} = 0.059$$

For 8 degrees of freedom, from Table A1 the value of $t_{95,8} = 2.310$, leading to a relative 95% confidence limit of $\pm \ 0.136$, and an absolute confidence limit for

$$\bar{L}_p \ of \pm 0.837 \ \mu m$$

Thus, rounding to significant figures gives

$$\bar{L}_p = 6.1 \pm 0.8 \ \mu m$$

In a similar way, the ferrite grain size is obtained from the number of ferrite grain boundaries, $n_{\alpha(i)}$, and the number of pearlite colony boundaries, $n_{p(i)}$, on each traverse, and the line fraction occupied by ferrite, i.e., $(1-P_{p(i)})$ as

$$N_{\alpha(i)} = \frac{(n_{\alpha(i)} + \frac{1}{2} n_{p(i)})}{(1 - P_{p(i)}) \times L} \tag{4.20}$$

Substitution of the values from Columns 3, 4 and 6 of Table 4.3 into eqn (4.20) then leads to the values given in Column 7.

Taking the mean value of $\bar{P}_p = 0.0176$ gives $(1 - \bar{P}_p) = 0.824$ and substituting the totals from Columns 4 and 6 into eqn (4.20) gives

$$\bar{N}_\alpha = \frac{(314 + 454/2)}{0.824 \times 0.88 \times 9} = 82.90 \ mm^{-1}$$

Thus, the mean ferrite grain size

$$\bar{L}_\alpha = \frac{1}{\bar{N}_\alpha} = 0.0121 \ mm$$

$$= 12.1 \ \mu m$$

In order to obtain the confidence limit of the mean, each traverse is considered as a measurement and from eqn (A.2), the standard deviation of $N_{\alpha(i)}$ is found as

$$s = 4.45 \text{ mm}^{-1}$$

Hence the standard error of $\overline{N}_\alpha$ is

$$S(\overline{N}_\alpha) = \frac{4.45}{\sqrt{9}} = 1.48 \text{ mm}^{-1}$$

The relative standard errors

$$\frac{S(\overline{L}_\alpha)}{\overline{L}_\alpha} = \frac{S(\overline{N}_\alpha)}{\overline{N}_\alpha} = 0.0179$$

For 8 degrees of freedom, the value of $t_{95,8} = 2.31$, leading to a relative 95% confidence limit of 0.0413 and an absolute confidence limit for $\overline{L}_\alpha$ of ± 0.5 μm.

Thus, rounding to significant figures gives

$$\overline{L}_\alpha = 12.1 \pm 0.5 \text{ μm}$$

Comments

1. The 95% confidence limits of $\overline{L}_\alpha$ of ± 0.5 μm are surprisingly low. The number of ferrite grains measured is $\Sigma n_\alpha(i) + \tfrac{1}{2}\Sigma nP_{(i)} = 314 + 354/2 = 491$, and eqn (2.10) indicates that for measurement of 491 grains in a single phase alloy the expected confidence limits would be

$$\frac{0.65}{\sqrt{491}} \times 12.1 \times 2 = \pm 0.7 \text{ μm}$$

    For the two phase alloy there is the additional uncertainty arising from the measurement of volume fraction, which will increase the confidence limits. The result thus shows that the presence of pearlite gives rise to more uniform ferrite grains than expected in a single phase alloy.
2. Care must be taken when differentiating between some grain boundaries and small pearlite colonies. This is particularly important if the sample has been heavily etched.

EXAMPLE 4.3.2 NUCLEATION AND GROWTH IN PHASE TRANSFORMATION (RECRYSTALLISATION)

Most phase transformations in metallurgical systems occur over a period of time by diffusion controlled nucleation and growth. Information about the nucleation and growth characteristics, as well as about the overall transformation rate can be obtained by carrying out a series of experiments with different annealing times and then quenching the specimens to 'freeze' the transforming structure for observation at room temperature.

The methodology described in Example 4.3.1 is then applied to each of the specimens and the results are analysed as described below for the recrystallisation of an aluminium alloy.

Method

Fig. 4.8 shows examples of the microstructures (observed optically using polarised light) on longitudinal sections of a high purity aluminium-5% magnesium alloy deformed in plane strain compression, quenched and then annealed for three different times in a salt bath at 380°C. The recrystallising grains are nearly equiaxed, whereas the deformed grains are highly elongated and in some cases show evidence of internal substructure. Distinguishing between the recrystallised and unrecrystallised grains (the two 'phases') is therefore relatively unambiguous until high fractions of recrystallisation have taken place, when the uncertainty increases. In the same way as for Example 4.3.1, grids of lines were superimposed on these micrographs and on other micrographs taken after the same and other annealing times. The intersections of the grids were used to count the number of points in recrystallised grains ($p_r(i)$) and the same traverse lines were used to count the number of boundaries separating recrystallised colonies of grains from unrecrystallised grains ($n_{u,r}(i)$) and the number of grain boundaries in recrystallised colonies ($n_{r,r}(i)$). This procedure is illustrated in Fig. 4.9.

Results

The results obtained for the 7 horizontal and 10 vertical lines in Fig. 4.9 are given in Columns 2, 4 and 6 of Table 4.4. In fact for this annealing time of 50 s measurements were made on 6 micrographs taken from different areas of the specimen, to give results on a total of 42 horizontal and 60 vertical lines. The totals given in Columns 2, 4 and 6 in Table 4.4 are from all the micrographs, not simply the one illustrated in Fig. 4.9. Thus in Column 2 a total of 133 points was counted in recrystallised grains for the total number of 420 points, whereas from the results of H1 to H7 and V1 to V10 24 points were counted in recrystallised grains for the total number of 7 × 10 for H and 10 × 7 for V points. Clearly, the same total points are counted on both traverses. Thus in Column 3, the fraction recrystallised on each horizontal line is the number in Column 2 divided by 10 and on each vertical line is the number in Column 2 divided by 7, but the mean volume fraction at the bottom of Column 3 of 0.316 is 133/420 and the confidence limit is given by substituting these values into eqn (2.3) for the standard error.

The numbers of migrating boundaries $n_{u,r}(i)$ counted on each of the horizontal and vertical traverses in Fig. 4.9 are given in Column 4 of Table 4.4. Again the totals are from the 6 micrographs. The length of the horizontal traverse line in Fig. 4.9 is 112.7 mm and of the vertical traverses is 76.3 mm. Thus at the magnification of ×150 the lengths (L) in the sample are 0.751 and 0.509 mm, respectively, leading to the values of number of migrating boundaries per unit length, $N_{u,r}$, shown in Column 5. It is notable

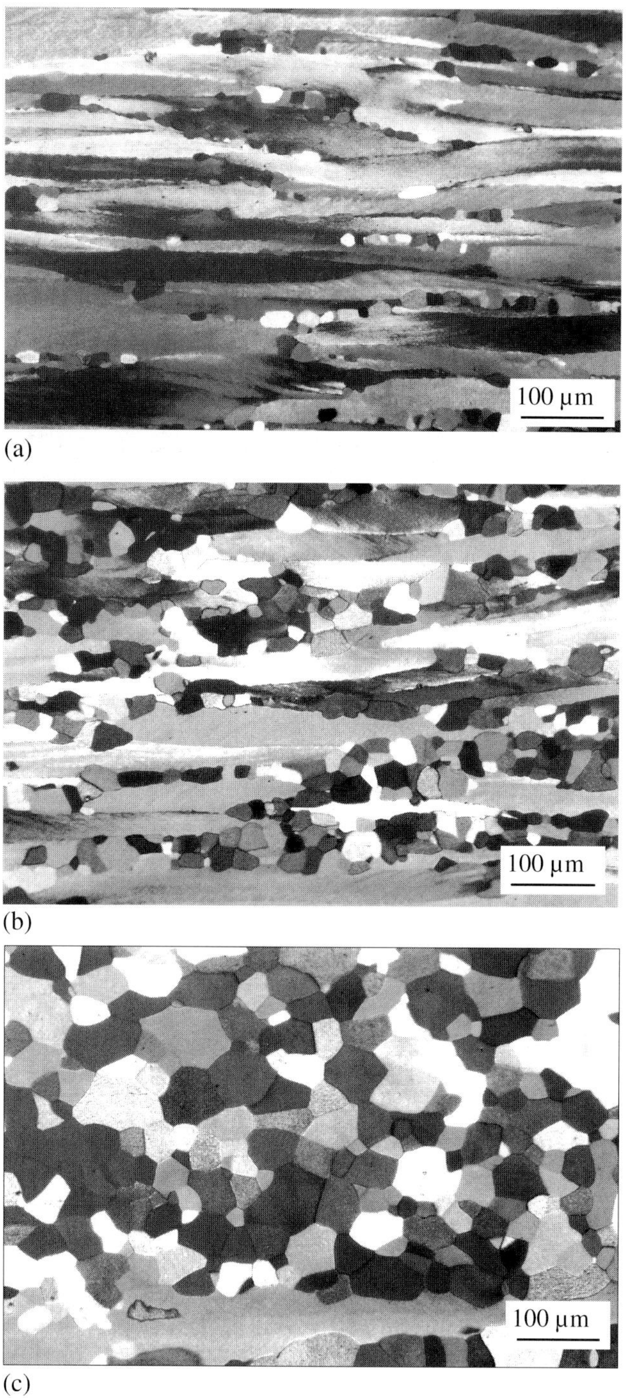

(a)

(b)

(c)

Fig. 4.8 Optical micrographs of longitudinal sections of high purity Al–5%Mg deformed in plane strain compression at a strain rate of 2.5 s^{-1} to a strain of 1.0 at 400°C, quenched and annealed in a salt bath at 380°C for (a) 25 s, (b) 50 s and (c) 120 s.

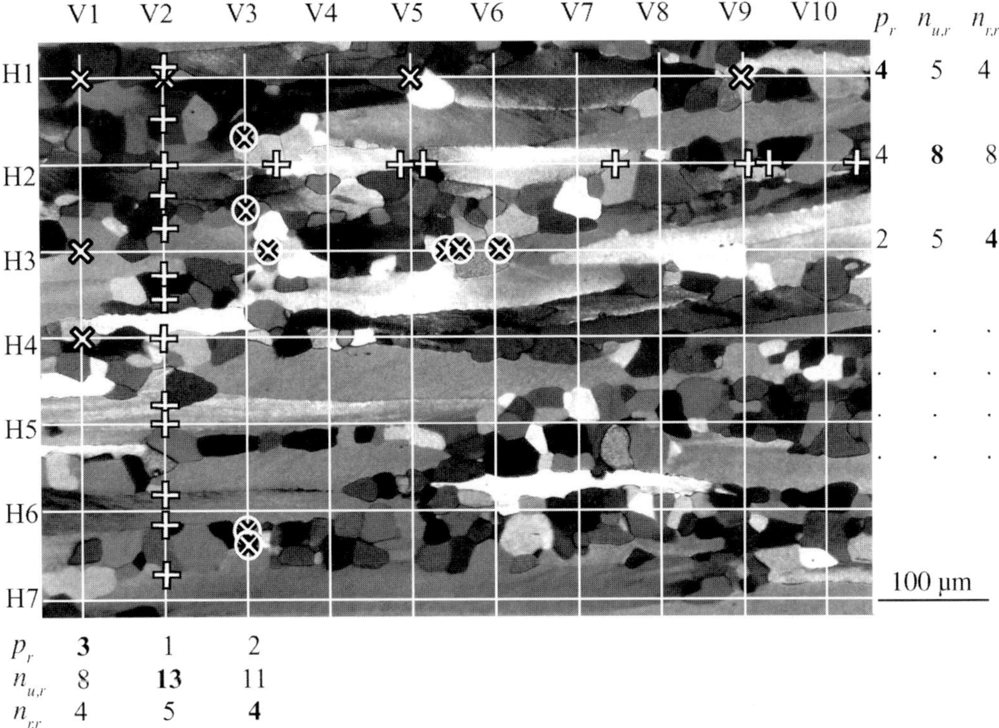

Fig. 4.9 Illustration of the procedure for counting points in recrystallised grains and the numbers of boundaries intersected along the *same* traverse lines on the micrograph shown in Fig. 4.8(b). Diagonal crosses in lines V1 and H1 indicate p_r, vertical crosses in H2 and V2 indicate $n_{u,r}$ and circled crosses in V3 and H3 indicate $n_{r,r}$.

that the values for the horizontal traverses (principal direction 1 of the plane strain deformation) are much lower than the values on the vertical traverses (principal direction 3 of the plane strain deformation), indicating that the colonies of recrystallised grains are elongated in the horizontal directions. This is expected from the qualitative observations from Fig. 4.8 that the recrystallised grains form preferentially along the boundaries of the elongated original grains. If it is assumed that the shape of the colonies of the recrystallising grains reflects the shape of the original grains deformed in plane strain compression, then from eqns (4.13) and (4.7) the area of migrating boundary per unit volume, $\bar{S}_{v(mig)}$, is obtained from the mean values $\bar{N}_{u,r}(1)$ from the horizontal traverses and $\bar{N}_{u,r}(3)$ from the vertical traverses as

$$\bar{S}_{v(mig)} = 0.429\bar{N}_{u,r}(1) + 0.571(\bar{N}_{u,r}(1)\bar{N}_{u,r}(3))^{\frac{1}{2}} + \bar{N}_{u,r}(3) \tag{4.21}$$

Substituting the mean values of $\bar{N}_{u,r}(1)$ from the horizontal traverses in Table 4.4 and the mean value of $\bar{N}_{u,r}(3)$ from the vertical traverses in Table 4.4 into eqn (4.21) leads to

Table 4.4 Results of point counting and counting boundaries illustrated in Fig. 4.9, for material annealed for 50 s.

(1) Traverse Number (i)	(2) Points in Rex. $p_r(i)$	(3) Fraction Rex. $X(i)$	(4) Mig. Boundaries $n_{u,r}(i)$	(5) No./Length. Mig. Boundaries $N_{u,r}(i)$ mm^{-1}	(6) Rex Bdies $n_{r,r}(i)$	(7) Rex. Grains $N_r(i)$ mm^{-1}
H1	4	0.40	5	6.66	4	21.6
H2	4	0.40	8	10.65	8	39.9
H3	2	0.20	5	6.66	5	43.3
H4	5	0.50	5	6.6	6	46.6
H5	6	0.60	1	1.33	17	46.6
H6	5	0.50	4	3.99	11	28.9
H7	1	0.10	3	3.99	0	20.00
⋮	⋮	⋮	⋮	⋮	⋮	⋮
Total	**133**		**173**		**384**	
Mean				**5.48**		
95%CL				**0.95**		
V1	3	0.43	8	2.129	4	3.105
V2	2	0.29	12	3.194	5	13.51
V3	1	0.14	11	2.928	3	5.822
V4	1	0.14	14	3.726	1	13.51
V5	2	0.29	10	2.661	1	4.890
V6	3	0.43	7	1.863	5	2.950
V7	2	0.29	6	1.597	3	3.493
V8	4	0.57	12	3.194	2	3.027
V9	3	0.43	13	3.460	5	4.813
V10	3	0.43	9	2.395	3	3.260
⋮	⋮	⋮	⋮	⋮	⋮	⋮
Total	**133**		**612**		**254**	
Mean		**0.316**		**20.04**		**5.281**
95%CL		**0.045**		**1.05**		**0.296**

Col 2 Experimental data from point counting,
Col 3 Obtained from Col 2 as Col 2 ÷ 10 for horizontal lines and ÷7 for vertical lines,
Col 4 Experimental data for number of migrating boundaries,
Col 5 Obtained as Col 4 ÷ traverse length,
Col 6 Experimental data for number of grain boundaries within recrystallised colonies,
Col 7 Obtained form Col 3, Col 4, Col 6, and eqn (4.22).

$$\overline{S}_{v(mig)} = 0.429 \times 5.48 + 0.571(5.48 \times 20.04)^{1/2} + 20.04$$

$$= 28.37 \text{ mm}^2/\text{mm}^3$$

$$\overline{\overline{S}}_{v(mig)} = 28.4 \pm 2.0 \text{ mm}^2/\text{mm}^3$$

The size of recrystallising grains is obtained in a similar way to the ferrite grain size in Example 4.3.1/eqn (4.20), by using boundary counts from both Columns 4 and 6 together with the fractions recrystallised from Column 3, to give the number of recrystallised grains per unit length,

$$N_r(i) = \frac{n_{r,r}(i) + \dfrac{1}{2} n_{u,r}(i)}{X(i)L} \tag{4.22}$$

shown in Column 7. Taking each of the 102 values from the 42 horizontal and 60 vertical lines as independent measurements, and applying the statistical procedures in the Appendix then leads to the overall mean value of $\overline{N}_r$ and its confidence limits shown at the bottom of Column 7 in Table 4.4.

From these values,

$$\overline{d}_r = \frac{1}{\overline{\overline{N}}_r} = 19.1 \,\mu\text{m}$$

Equating the relative 95% confidence limits, from Column 7

$$\frac{2S(\overline{d}_r)}{\overline{d}_r} = \frac{2S(\overline{N}_r)}{\overline{\overline{N}}_r} = \frac{2.96}{52.36} = 0.057$$

Thus, $\overline{d}_r = 19.1 \pm 1.1 \,\mu\text{m}$.

Growth Rate

Cahn and Hagel (1960) derived a general relationship for the average growth rate of transforming regions at any time during transformation as

$$\overline{G} = \frac{1}{\overline{S}_{v(mig)}} \times \frac{dX}{dt} \tag{4.23}$$

where the migrating boundary area per unit volume, $\overline{S}_{v(mig)}$, and the rate of transformation, dX/dt, are the values at each time of interest.

In order to obtain the growth rates, it is therefore necessary to have the results for other annealing times. These have been obtained in the same way as for Table 4.4, so only the mean values and confidence limits of the essential parameters are shown in Table 4.5.

Table 4.5 Data obtained by quantitative measurement of the microstructures of high purity A1-5% Mg specimens annealed in a salt bath for various times after plane strain compression testing and quenching to room temperature.

(1)	(2)	(3)	(4)	(5)	(6)	(7)
Annealing Time, s	Equivalent Time, s	$\overline{X}$	$\overline{S}_{v(mig)}$, mm^{-1}	dX/dt s^{-1} x1000	$\overline{G}$ µm/s	$\overline{d}_r$ µm
0	5.38	0.128 ± 0.033	21.6 ± 1.5	26.3	1.22 ± 0.08	15.4 ± 1.2
25	5.78	0.119 ± 0.032	20.9 ± 1.4	26.7	1.28 ± 0.08	15.3 ± 1.2
50	9.95	0.316 ± 0.045	28.4 ± 2.0	21.9	0.77 ± 0.05	19.1 ± 1.1
80	26.34	0.424 ± 0.048	28.1 ± 1.9	20.3	0.72 ± 0.05	25.3 ± 1.3
100	42.78	0.757 ± 0.042	14.6 ± 1.1	9.00	0.62 ± 0.05	37.5 ± 1.6
120	61.67	0.931 ± 0.025	7.02 ± 0.71	2.65	0.38 ± 0.04	46.8 ± 1.7
240	181.67	0.989 ± 0.010	1.94 ± 0.30	0.45	0.23 ± 0.04	47.8 ± 1.7

Col 1 Experiential times of annealing,

Col 2 Equivalent times of isothermal annealing at 380°C from Table 4.6,

Col 3 Fractions recrystallised from experimental point counts,

Col 4 Migrating boundary areas per unit volume from measurements of $N_{u,r}$ and eqn (4.21),

Col 5 Recrystallisation rates from Fig. 4.11 and eqn (4.27),

Col 6 Mean growth rate of recrystallising colonies from Col 4, Col 5 and eqn (4.23),

Col 7 Mean size of recrystallising grains from experimental boundary counts and eqn (4.22).

The determination of dX/dt requires accurate knowledge of the time at the annealing temperature (in this case 380°C) in order to differentiate the transformation curve. If the times are long, then the true isothermal annealing time will approximate closely to the measured annealing time, but when the times are short, (as in the present example) it is necessary to make allowance for any delay in quenching after deformation, and for the time taken to heat the specimens to the annealing temperature.

This can be done approximately by recognising that the stored energy in the deformed structure is determined at the end of deformation. The kinetics of the subsequent recrystallisation are thermally activated and so depend on the instantaneous value of temperature through the activation energy for recrystallisation, Q_{rex}.

Thus an increment of time, Δt_x, at any temperature, T_x, can be equated to an equivalent increment of time, Δt_a, at the annealing temperature, T_a, as

$$\Delta t_a = \Delta t_x \exp \frac{Q_{rex}}{R} \left(\frac{1}{T_a} - \frac{1}{T_x} \right) \tag{4.24}$$

where R is the gas constant ($8.314 \, J \, mol^{-1} \, K^{-1}$) and T_a and T_x are the absolute temperatures. When temperature is changing, $(1/T)_x$, is the mean value over the time interval Δt_x.

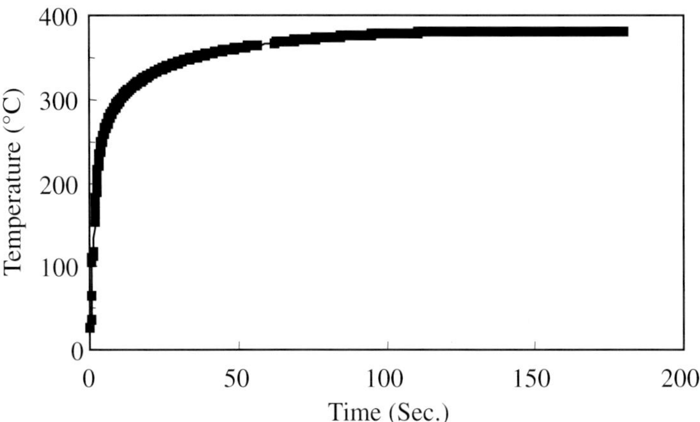

Fig. 4.10 Temperature versus time curve logged from a 1.5 mm diameter K type thermocouple embedded in a sample annealed in a salt bath at 380°C.

In the present example, specimens were deformed at 400°C and then water quenched. Once the water contacted the specimen, cooling was extremely rapid, but there was a delay of 1.8 s between the end of deformation and the quench. The value of Q_{rex} for Al-5%Mg has been reported as 200 kJ mol^{-1} (Wells et al. (1998)). Taking this value and substituting in eqn (4.24) gives an equivalent increment of annealing time at 380°C of

$$\Delta t_{380} = 1.8 \exp \frac{200000}{8.314} (\frac{1}{653} - \frac{1}{673})$$

i.e.,

$$\Delta t_{380} = 5.38 \text{ s}$$

After quenching, small samples were cut from the tested specimens for annealing in a salt bath. Nevertheless, the heating rate was surprisingly slow, as shown in Fig. 4.10.

In order to calculate the equivalent annealing times at 380°C, the curve can be considered as a sequence of isothermal steps over the time increments, Δt_x. In selecting the increments during heating it is convenient to ensure that some of them finish at the annealing times at which samples were removed and quenched. In the present example these times are 25, 50, 80 and 100 s, giving the selected times shown in Column 1 of Table 4.6, for which the temperatures read from Fig. 4.10 are shown in Column 2. Subtracting the times sequentially gives the time increments shown in Column 3. Over these time increments the mean reciprocal absolute temperature

Table 4.6 Analysis of data from the heating curve in Fig. 4.10 to obtain equivalent isothermal annealing times at 380°C.

(1)	(2)	(3)	(4)	(5)	(6)	(7)
t_x, s	T_x °C	Δt_x, s	$(1/T)_x \mathrm{K}^{-1} \times 10^3$	Δt_{380}, s	$\Sigma \Delta t_{380}$, s	Total t_{380}, s
2.6	200	2.6		5.38	-	
3.3	225	0.7	2.06	2.05×10^{-6}	-	
4.4	250	1.1	1.96	3.66 c 10^{-5}	-	
6.6	275	2.2	1.87	0.001	0.00 + 5.38	
11.0	300	4.4	1.79	0.01	0.01 + 5.38	
18.4	325	7.4	1.71	0.10	0.11 + 5.38	
25.0	333	6.6	1.66	0.29	0.40 + 5.38	5.78
41	352	11.8	1.63	1.68	2.08 + 5.38	
50.0	364	13.2	1.59	2.48	4.56 + 5.38	9.94
80.0	375	30.0	1.56	16.39	21.0 + 5.38	26.3
100	378	20.0	1.54	16.40	37.4 + 5.38	42.8
120	380	20.0	1.53	18.90	56.3 + 5.38	61.7
240	380	120		120	176.3 + 5.38	181.7

Col 1 Times selected for discretising the heating curve,
Col 2 Temperatures at the selected times in Col 1,
Col 3 Time increments from Col 1,
Col 4 Mean reciprocal temperature for each time increment from Col 3 and eqn (4.25),
Col 5 Equivalent time increment at 380°C from Col 4 and eqn (4.24),
Col 6 Total equivalent time at 380°C during heating, from Col 5, and quenching delay,
Col 7 Equivalent times at 380°C from Col 6 for the experimental annealing times.

$$\left(\frac{\overline{1}}{T}\right)_x = \frac{1}{2}\left(\frac{1}{T_x} + \frac{1}{T_{x-1}}\right) \tag{4.25}$$

Substitution in eqn (4.25) gives the values in Column 4 and hence from eqn (4.24) the equivalent time increments in Column 5. It can be seen that below 300°C these time increments are negligible. To obtain the equivalent isothermal annealing times at which specimens were examined, the increments are summed in Column 6 and added to the increment from the quenching delay and, for times longer than 100 s, are added to the true isothermal time interval, to given the total equivalent times in Column 7.

These times have been used to plot the recrystallisation curve for 380°C, shown in Fig. 4.11. As considered in Chapter 2, the best method of smoothing the data to obtain the transformation rate dX/dt is to fit the Avrami Equation

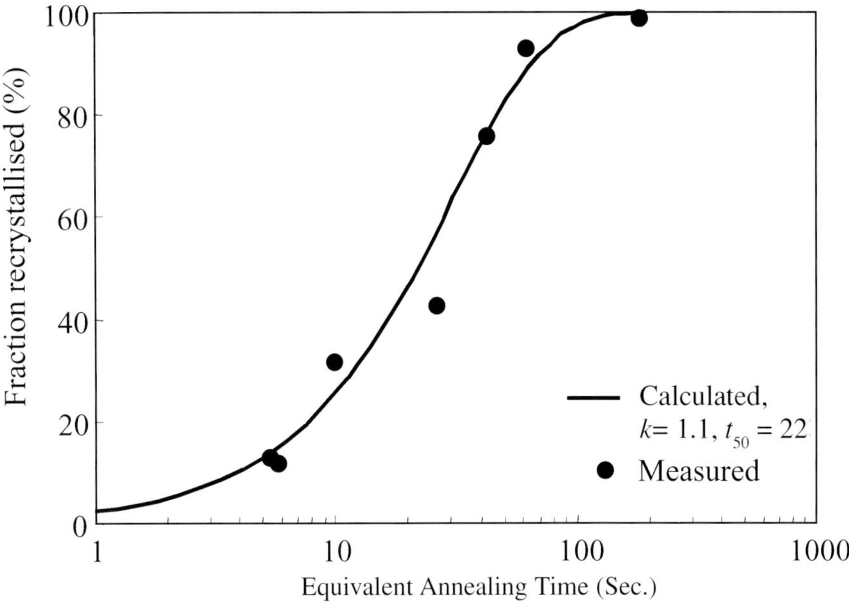

Fig. 4.11 Recrystallised fraction as a function of isothermal time of annealing at 380°C.

$$X = 1 - \exp - 0.693 \left(\frac{t}{t_{50}} \right)^k \tag{4.26}$$

where t_{50} is the time for 50% transformation ($X = 0.5$) and k is a constant. Both values are found by plotting $\ln 1/(1 - X)$ versus t (logarithmic scales) to obtain a straight line of slope k and intercept of t_{50} at $\ln 1/0.5 = -0.693$. Fig. 4.12 shows the plot for the present data. It can be seen that, except for the point near the end of transformation, a good straight line is obtained, giving values of $k = 1.1$ and $t_{50} = 22$ s.

Differentiation of eqn (4.26) gives

$$\frac{dX}{dt} = 0.693k \frac{(t)^{k-1}}{(t_{50})^k} (1 - X) \tag{4.27}$$

and substituting the values of k and t_{50}, and the values of X at the equivalent times in Column (2) of Table 4.5 gives the values of dX/dt in Column 5 of Table 4.5, and hence from eqn (4.23) the values of $\overline{G}$ in Column 6. These values are plotted against time (logarithmic scales) in Fig. 4.13, from which it can be seen that the average growth rate decreases with time as

$$\overline{G} = \overline{G}_o t^{-n} \tag{4.28}$$

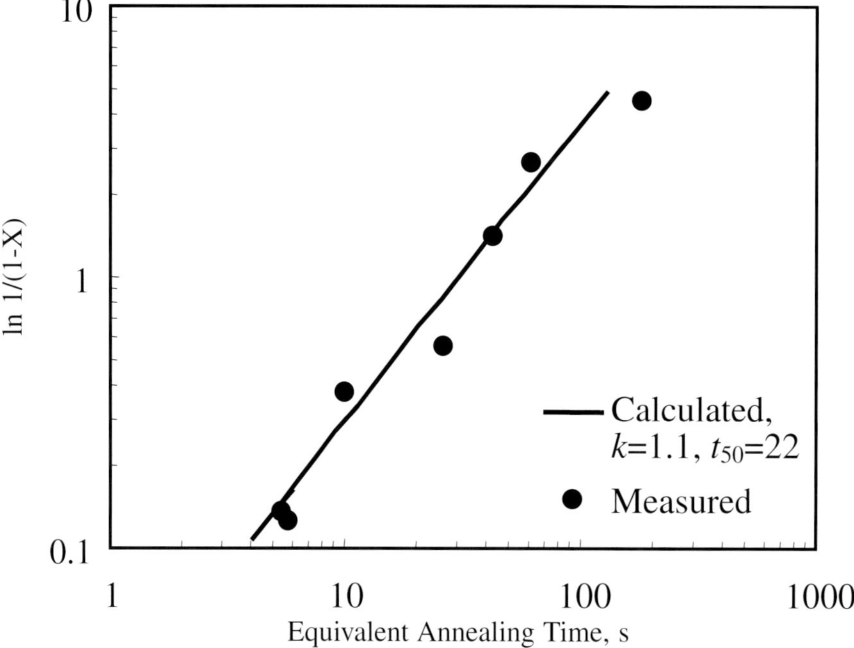

Fig. 4.12 Avrami plot of recrystallisation for isothermal annealing at 380°C.

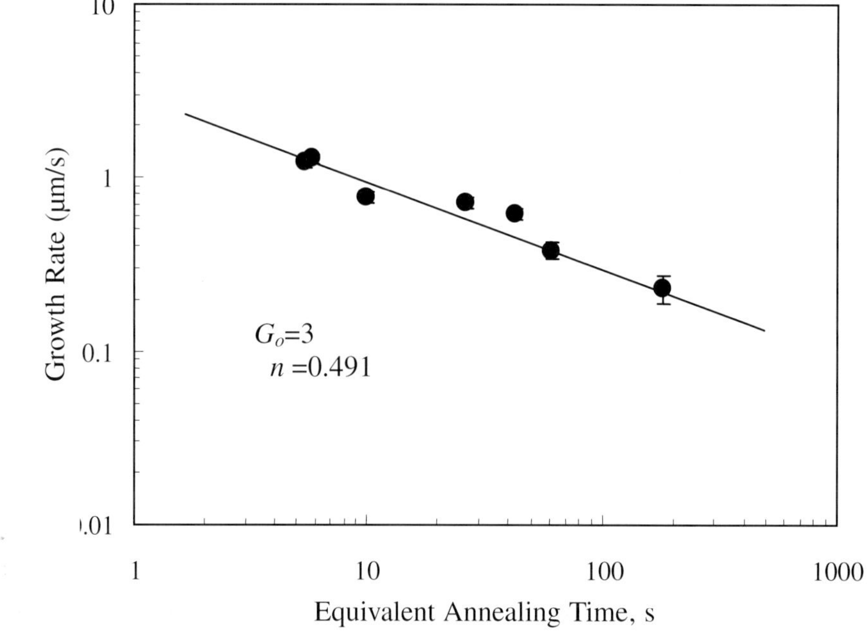

Fig. 4.13 Dependence of the mean rate of growth of recrystallising grains on annealing time.

during the transformation, with values of $\bar{G}_0 = 3$ µm/s from the value of $\bar{G}$ when $t = 1$ s, and $n = 0.49$ from the slope of the line.

Nucleation

It is more difficult to obtain quantitative data about nucleation, but at any instant the fraction recrystallisation is related to the number per unit volume of growing grains, N_V, and their mean size, $\bar{d}_r$, as

$$X = AN_V \bar{d}_r^{\,3} \tag{4.29}$$

where A is a geometrical constant, which depends on the size distribution and shape of the growing grains. If these do not change during recrystallisation,

$$X = \frac{N_V}{N_{V(rex)}} \left(\frac{\bar{d}_r}{\bar{d}_{rex}} \right)^3 \tag{4.30}$$

where $N_{V(rex)}$ is the number of grains per unit volume of mean size $\bar{d}_{rex}$ when $X = 1$. If coarsening of grains in the recrystallised regions is negligible during the time of recrystallisation, i.e., grain coarsening (growth) after recrystallisation is slow, then N_V is a measure of the density of nucleation sites. The commonly made assumption of 'site saturation', i.e., all nuclei being present at a very early stage of transformation leads to a constant value of N_V and hence the expectation from eqn (4.30) that $\bar{d}_r/\bar{d}_{rex} \propto X^{1/3}$. This relationship is shown as the broken line in Fig. 4.14, in which the data points are from Column 7 of Table 4.5. It is clear that the assumption of site saturation is not valid for the conditions of deformation and annealing in the present example.

Comments

1. Generally the number of grain boundaries in unrecrystallised grains, $n_{u,u}(i)$, can be counted at the same time as $n_{u,r}(i)$ and $n_{r,r}(i)$ and these measurements can give some indication of the preferred nucleation sites (Orsetti-Rossi and Sellars (1997)). In the present example, the coarse original grain size ($d_0 = 84 \pm 4$ µm) and large strain meant that too few grain boundaries were encountered along the experimental traverses for the results to be significant.
2. A more detailed analysis of the results from the type measurements made in the present example can provide more information about the characteristics of nucleation and growth than considered above (Orsetti-Rossi and Sellars (1997)).
3. In the present example the corrections required to obtain the equivalent isothermal annealing time from the time in the salt bath are rather large. An accurate value of Q_{rex} is therefore required. The corrections also indicate that 380°C is about the highest annealing temperature for which deformation, quenching and subsequent annealing can be used to study recrystallisation in the experimental Al-5%Mg. Above

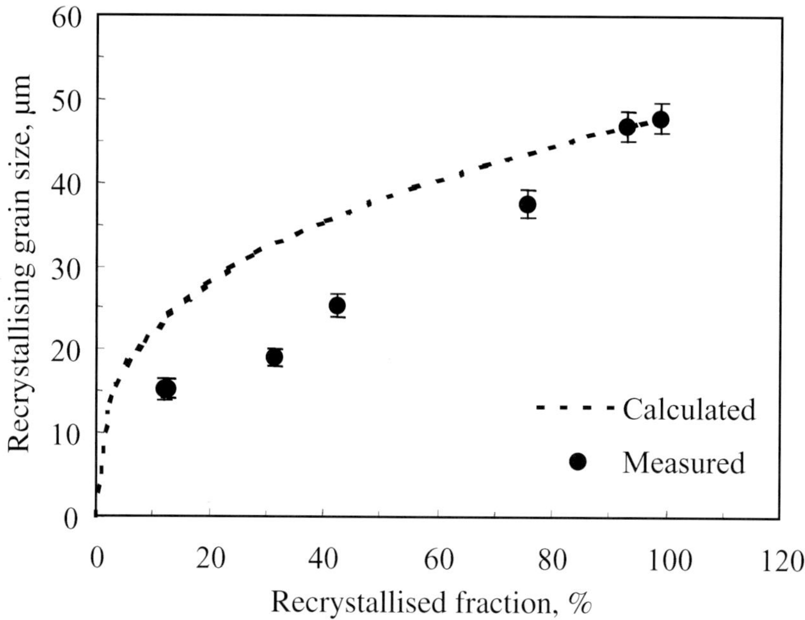

Fig. 4.14 Dependence of size of the recrystallising grains on the volume fraction recrystallised compared with calculated values assuming site saturation.

this temperature the annealing must be carried out after deformation and before quenching to avoid excessive corrections for equivalent isothermal times. This experimental procedure has the advantage that it avoids any possibility of secondary effects, e.g., precipitation phenomena, interfering with the recrystallisation results. This procedure may therefore be essential for some alloys, but it has the disadvantage that a new test specimen must be deformed for each annealing time.

4.   As an alternative to analysing the temperature versus time curve during annealing in terms of a series of isothermal steps, as used in the example, a polynomial or other form of curve fitting may be used. In this case a good fit is only required for about 100°C below the annealing temperature.

5. Size Distributions of Second Phase Particles

As discussed in Chapter 2, particle sizes may be in a size range where they are most conveniently measured on plane sections by optical or scanning electron microscopy, or they may have to be measured from transmission electron micrographs of extraction replicas or thin foils. In all cases the method of measurement is the same, but the analysis of the data to obtain the size distribution in the volume, which is frequently the one of interest, is different.

It is usual to assume that the particles are spheroidal in the volume so that they can be considered in terms of equivalent sphere diameters or equivalent circle diameters in sections. Analyses for other shapes such as circular plates or rods have been derived (Fullman, 1953, De Hoff and Rhines, 1961, Underwood, 1970) but are more complex and will not be considered here.

5.1 METHOD OF MEASUREMENT

It is normal practice to measure particle sizes from micrographs rather than directly in the microscope and it is essential that the magnification is accurately calibrated.

For near-spherical particles, measurements are made of the equivalent circle diameters of all particles seen in the area of observation. The basic method to do this "by hand" is to compare the image of each particle with a series of circles, of say, 1 to 15 mm diameter at intervals of 1 mm, to count the particle in the right size range in order to generate the size distribution in the form of a histogram.

The simplest, but most tedious, way is to use a series of circles printed on, or punched or drilled out of thin transparent material, e.g. Perspex. Having the circle as a hole is advantageous because it facilitates marking each particle image as it is measured, to ensure that no particles are missed and that no particles are measured twice. Fig. 5.1 illustrates how the correct size is selected when particles are not perfectly spherical. In Fig. 5.1(a) the 14 mm diameter circle is clearly of smaller area than the particle. In Fig. 5.1(b) the 15 mm diameter circle intersects the particle with slightly less particle area outside the circle than circle area is outside the particle. In Fig. 5.1(c) the 16 mm diameter circle is clearly of larger area than the particle. The equivalent circle diameter of the particle image is therefore correctly placed in the size range 14 to 15 mm.

This method may be speeded up by using a Zeiss particle size analyser, in which the image of a particle in a micrograph printed on thin paper is centred within a diaphragm.

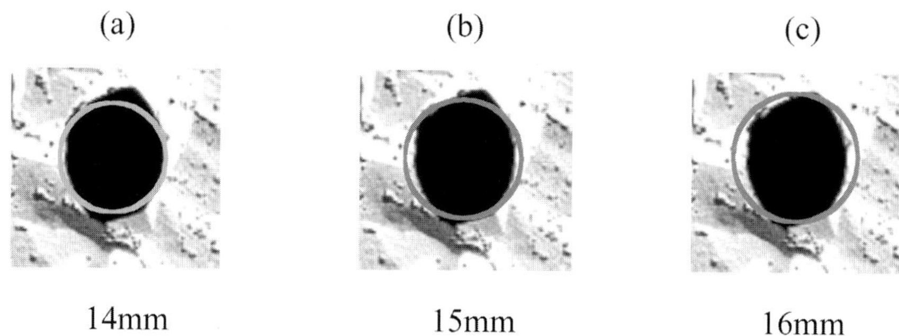

Fig. 5.1 Comparison of a particle with circles of increasing size.

The micrograph is illuminated from behind, so the particle appears as a silhouette and the diaphragm is adjusted in diameter until the image is similar to that in Fig. 5.1(b). When a lever switch is pressed, the size interval is registered in a counter and a small hole is punched in the centre of the particle image. Light shining through the hole clearly indicates that a particle has been measured. With practice, several hundred particles can be measured in less than an hour.

If an automatic image analysis system is available, this provides the quickest method of measurement providing that the particle images are not confused by matrix contrast. In this case the areas of the particle images are measured directly in terms of numbers of pixels and are converted into equivalent circle diameters by the software, to present the particle size distribution as a histogram.

5.2 ANALYSIS OF PLANAR SECTIONS

When measurements are made on planar sections by optical or scanning electron microscopy, the sections of particles appear as circles in the micrograph. However, the number of circles per unit area (N_A) arising from (N_V) spheres per unit volume of diameter, D is

$$N_A = D \cdot N_V \tag{5.1}$$

i.e. larger spheres are more likely to be intersected by the plane of polish than smaller ones. Furthermore, a sphere may be sectioned anywhere in its diameter, to give circles of diameter (d) ranging in size from O to D. The shape of the size distribution of circles in the section therefore differs markedly from the distribution of spheres in the volume.

Scheil (1935) developed a method of converting the observed distribution of circles into the volume distribution of spheres. His method recognised that the largest circle diameter could only arise by sectioning the largest spheres. The probability of observing circles in this largest size group could then be calculated and the residual probability distributed to the smaller size groups of circles. Circles arising from the next smallest size group of spheres were calculated etc. This procedure enables the size distribution of spheres in the volume to be derived from the observations of the size distributions of circles on the plane of polish, but the successive subtractions involved led to large statistical uncertainties in the numbers of particles in the smallest size groups of spheres. Today the procedure used is often referred to as 'Scheil analysis', even though the methods applied are those developed by Schwartz (1934) and Saltykov (1952), or by Woodhead (1980). These methods develop a matrix of coefficients $(a(i, j))$ for the number of circles in size group (i) arising from spheres in size group (j) using probability distributions for sectioning randomly distributed spheres of sizes in k equal size groups. By matrix inversion both methods provide sets of coefficients $(\alpha(j,i))$ to derive the number of spheres per unit volume in size group j from the numbers of circles in size groups i, where $i = 1$ to k, $k\Delta$ is the maximum size of circles (and spheres), and Δ is the size interval used in the histograms. The coefficients are listed in Tables 5.1 and 5.2, respectively.

The coefficients differ because, in deriving the coefficients for the Schartz-Saltykov analysis, it was assumed that the spheres are of discrete sizes, $j\Delta$, where $j = 1$ to k, whereas Woodhead considered that the spheres arise from a histogram with sizes over the range Δ, i.e. from $(j\text{-}i)\,\Delta$ to $j\Delta$, with means of $(j{-}\tfrac{1}{2})\,\Delta$ at the centre of each interval. This leads to important differences in interpretation of the results of the analysis, even though the coefficients are applied in the same way, i.e.

$$N_V(j) = \frac{1}{\Delta}\,\{\alpha(j, j)\,N_A(j) + \alpha(j, j{+}1)\,N_A(j + 1) + \dots\dots + \alpha(j, k)\,N_A(k)\} \tag{5.2}$$

EXAMPLE 5.2.1 – SCHWARTZ-SALTYKOV AND WOODHEAD ANALYSES

In order to illustrate the application of the Schwartz-Saltykov and Woodhead analyses, a frequency distribution of circle sizes $(N_A(i))$ shown by the solid histogram in Fig. 5.2 has been derived from a known normal distribution of spheres $(N_V(j))$ of mean size $\bar{D}_V = 30$ µm, with a size interval $\Delta = 6$ µm, shown as the broken line histogram in Fig. 5.2. This clearly illustrates the difference in shape of the area and volume size distributions.

The analysis also leads to the figures for the number of circles per unit area observed in each size group, given in Column 3 of Table 5.3. This column is equivalent to experimental data derived from numbers (n_i) of circles in each size group obtained from micrographs of area (A), taken at magnification (M), i.e.

$$N_A(i) = \frac{n_i\,M^2}{A} \tag{5.3}$$

and

Table 5.1 Coefficients for the Schwartz-Saltykov method of calculating particle size distributions.

	Coefficients, $\alpha(ji)$														
	$(N_A)_1$	$(N_A)_2$	$(N_A)_3$	$(N_A)_4$	$(N_A)_5$	$(N_A)_6$	$(N_A)_7$	$(N_A)_8$	$(N_A)_9$	$(N_A)_{10}$	$(N_A)_{11}$	$(N_A)_{12}$	$(N_A)_{13}$	$(N_A)_{14}$	$(N_A)_{15}$
$(N_V)_1$	+1.0000	0.1547	0.03602	0.0130	0.0061	0.0033	0.0020	0.0013	0.0009	0.0006	0.0005	0.0004	0.0003	0.0002	0.0001
$(N_V)_2$		+0.5774	0.1529	0.0420	0.0171	0.0087	0.0051	0.0031	0.0021	0.0015	0.0010	0.0009	0.0006	0.0006	0.0004
$(N_V)_3$			+0.4472	0.1382	0.0408	0.0178	0.0093	0.0057	0.0037	0.0026	0.0018	0.0013	0.0010	0.0007	0.0007
$(N_V)_4$				+0.3779	0.1260	0.0386	0.0174	0.0095	0.0058	0.0038	0.0027	0.0020	0.0016	0.0012	0.0009
$(N_V)_5$					+0.3333	0.1161	0.0366	0.0168	0.0094	0.0059	0.0040	0.0028	0.0021	0.0016	0.0013
$(N_V)_6$						+0.3015	0.1081	0.0346	0.0163	0.0091	0.0058	0.0041	0.0028	0.0022	0.0016
$(N_V)_7$							+0.2773	0.1016	0.0329	0.0155	0.0090	0.0057	0.0040	0.0029	0.0022
$(N_V)_8$								+0.2582	0.0961	0.0319	0.0151	0.0088	0.0056	0.0039	0.0028
$(N_V)_9$									+0.2425	0.0913	0.0301	0.0146	0.0085	0.0055	0.0039
$(N_V)_{10}$										+0.2294	0.0872	0.0290	0.0140	0.0083	0.0054
$(N_V)_{11}$											+0.2182	0.0836	0.0280	0.0136	0.0080
$(N_V)_{12}$												+0.2085	0.0804	0.0270	0.01321
$(N_V)_{13}$													+0.2000	0.0776	0.0261
$(N_V)_{14}$														+0.1925	0.0750
$(N_V)_{15}$															+0.1857
N_V	+1.000	+0.4227	+0.2583	+0.1847	+0.1433	+0.1170	+0.0988	+0.0856	+0.0753	+0.0672	+0.0610	+0.0553	+0.0511	+0.0472	+0.0441

NB: All Unsigned Coefficients are Negative.

Table 5.2 Coefficients for the Woodhead method of calculating particle size distributions.

	Coefficients, $\alpha(ji)$														
	$(N_A)_1$	$(N_A)_2$	$(N_A)_3$	$(N_A)_4$	$(N_A)_5$	$(N_A)_6$	$(N_A)_7$	$(N_A)_8$	$(N_A)_9$	$(N_A)_{10}$	$(N_A)_{11}$	$(N_A)_{12}$	$(N_A)_{13}$	$(N_A)_{14}$	$(N_A)_{15}$
$(N_V)_1$	+2.00	−0.7944	+0.1801	−0.0665	+0.0143	−0.0075	+0.0005	−0.0013	−0.0004	−0.0005	−0.0003	−0.0002	−0.0002	−0.0001	−0.0001
$(N_V)_2$		+0.9315	−0.5595	+0.1112	−0.0538	+0.0067	−0.0077	−0.0011	−0.0020	−0.0010	−0.0009	−0.0006	−0.0005	−0.0004	0.0003
$(N_V)_3$			+0.6997	−0.4710	+0.0896	−0.0491	+0.0041	−0.0080	−0.0018	−0.0024	−0.0014	−0.0012	−0.0008	−0.0007	−0.0005
$(N_V)_4$				+0.5840	−0.4150	+0.0773	−0.0454	+0.0028	−0.0079	−0.0022	−0.0026	−0.0016	−0.0013	−0.0010	−0.0008
$(N_V)_5$					+0.5116	−0.3752	+0.0691	−0.0425	+0.0020	−0.0078	−0.0024	−0.0027	−0.0017	−0.0014	−0.0011
$(N_V)_6$						+0.4608	−0.3451	+0.0631	−0.0400	+0.0014	−0.0076	−0.0025	−0.0027	−0.0018	−0.0015
$(N_V)_7$							+0.4226	−0.3212	+0.0584	−0.0379	+0.0011	−0.0074	−0.0025	−0.0027	−0.0018
$(N_V)_8$								+0.3926	−0.3017	+0.0547	−0.0361	+0.0008	−0.0072	−0.0025	−0.0027
$(N_V)_9$									+0.3682	−0.2853	+0.0516	−0.0345	+0.0006	−0.0070	−0.0025
$(N_V)_{10}$										+0.3478	−0.2714	+0.0489	−0.0332	+0.0005	−0.0068
$(N_V)_{11}$											+0.3305	−0.2593	+0.0467	−0.0319	+0.0003
$(N_V)_{12}$												+0.3155	−0.2487	+0.0447	−0.0308
$(N_V)_{13}$													+0.3024	−0.2393	+0.0430
$(N_V)_{14}$														+0.2908	−0.2309
$(N_V)_{15}$															+0.2805

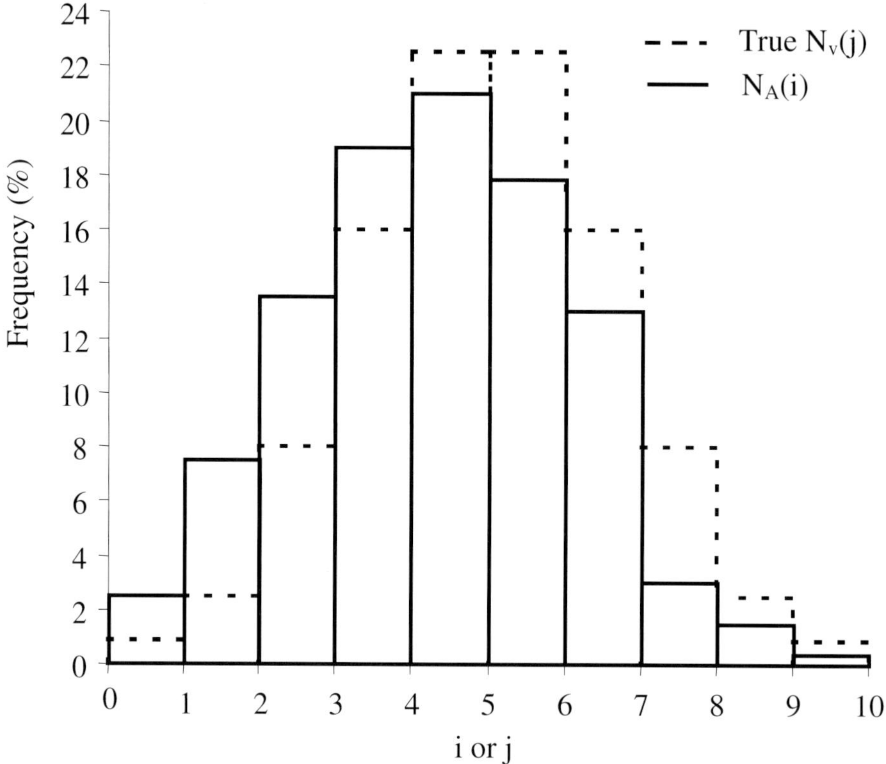

Fig. 5.2 Histogram of the area size distribution of circles on a plane ($N_A(i)$) derived from a normal distribution of spheres in the volume ($N_V(j)$).

$$\Delta = \frac{\Delta(\text{micrograph})}{M} \tag{5.4}$$

Columns 4 and 5 in Table 5.3 show the volume distributions derived by the Schwartz-Saltykov and Woodhead methods of analysis, respectively. The way in which the numbers are obtained by applying eqn 5.2 is illustrated below for group 5.

Schwartz-Saltykov: (see Table 5.1 for the coefficients for group 5).

$$N_V(5) = \frac{1}{6\times10^{-3}} \{0.3333 \times 63.21 - 0.1161 \times 53.47 - 0.0366 \times 33.55$$
$$- 0.0168 \times 15.30 - 0.0094 \times 4.952 - 0.0059 \times 0.990\} = 2221$$

Woodhead: (see Table 5.2 for the coefficients for group 5)

Table 5.3 Volume particle size distributions derived from the area distributions of circle sizes in Fig. 5.2 using Schwartz-Saltykov and Woodhead analyses.

(1)	(2)	(3)	(4)	(5)
Group Number	Upper Diameter (μm)	N_A (mm^{-2})	N_V (mm^{-3}) S-S	N_V (mm^{-3}) W
1	6	7.218	140.5	54
2	12	22.80	461.6	250
3	18	40.66	1058	770
4	24	57.05	1796	1616
5	30	63.21	2221	2324
6	26	53.47	1979	2302
7	42	33.55	1262	1586
8	48	15.30	573.8	761
9	54	4.952	185.1	257
10	60	0.990	37.84	57
		299.200	9714.84	9977

Col 2 Upper diameter of circles in histogram of particles numbers observed on sections.
Col 3 Number per unit area of circles observed in each size group; equivalent to results of experimental measurements.
Col 4 Number per unit volume of spheres computed from Col 3 using the Schwartz–Saltykov coefficients, Table 5.1, in eqn (5.2).
Col 5 Number per unit volume of spheres computed from Col 3 using the Woodhead coefficients, Table 5.2, in eqn (5.2).

$$N_V(5) = \frac{1}{6 \times 10^{-3}} \{0.5116 \times 63.21 - 0.375 \times 53.47 + 0.0691 \times 33.55$$
$$- 0.0425 \times 15.30 + 0.0020 \times 4.952 - 0.0078 \times 0.990\} = 2324$$

From the numbers per unit volume, the frequency distributions are simply obtained as

$$F_v(i) = \frac{N_v(i)}{\Sigma N_v(i)} \tag{5.5}$$

and histograms are plotted from the Schwartz–Saltykov analysis in Fig. 5.3 and from the Woodhead analysis in Fig. 5.4. In this example these histograms can be compared with the true volume size distribution and histogram, shown by the broken lines. (These are, of course, not known for experimental measurements of real particle size distributions).

From Fig. 5.3 it is clear that the histogram plotted from the results of the Schwartz–Saltykov analysis does not correspond with the true size distribution. However the upper end of each size interval (marked by a cross) is close to the distribution curve. This is to be expected from the derivation of the Schwartz-Saltykov coefficients, which assumes

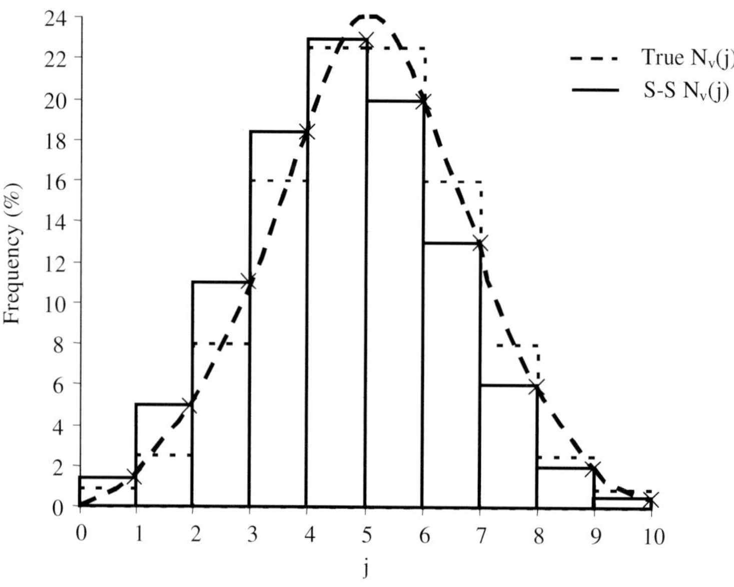

Fig. 5.3 Histogram of the volume size distribution of spheres derived by the Schwartz–Saltykov analysis (solid lines) compared with the true size distribution, (broken lines).

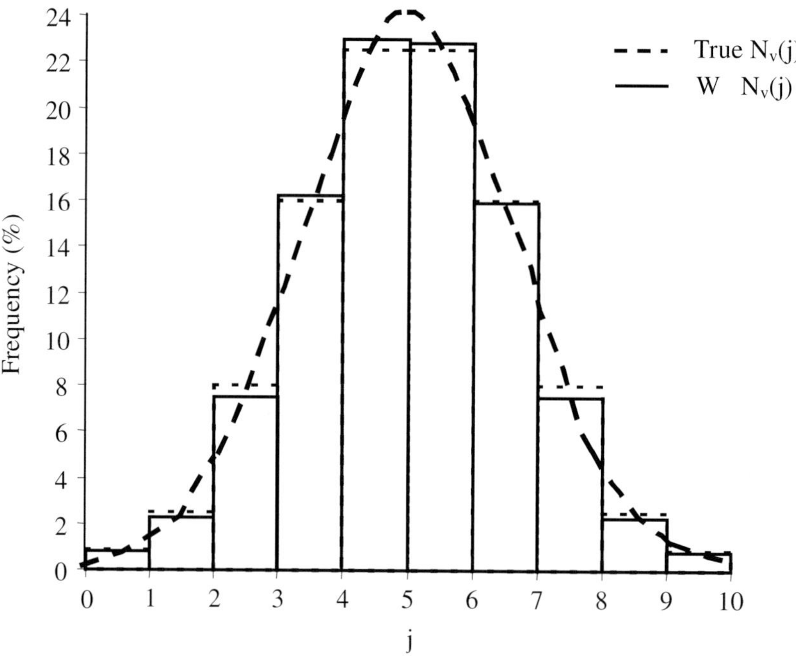

Fig. 5.4 Histogram of the volume size distribution of spheres derived by the Woodhead analysis (solid lines) compared with the true size distribution, (broken lines).

Table 5.4 Results of computation of particle size in the volume and its confidence limits from the data in Table 5.3.

	S-S	W
$\sum N_v(j) \cdot (j-\frac{1}{2}) \Delta$		299.07 mm
$\sum N_v(j) \cdot (j) \Delta$	299.24 mm	
$\sum N_v(j)$	9716	9977
$\overline{D}_V$	3.08×10^{-2} mm	3.00×10^{-2} mm
$\sum N_v(j) \cdot \left((j-\frac{1}{2})\Delta \right)^2$		9.915 mm^2
$\sum N_v(j) \cdot \left((j) \Delta \right)^2$	10.238 mm^2	
$\overline{D}_V^{\,2} \sum N_v(j)$	9.216 mm^2	8.967 mm^2
S	10.25×10^{-3} mm	9.74×10^{-3} mm
$S(\overline{D}_V)$ *	4.58×10^{-4} mm	4.36×10^{-4} mm
Thus $\overline{D}_V \pm 95\%$ CL	30.8 ± 0.9 µm	30.0 ± 0.9 µm

*Assuming that 500 particle images are measured, i.e. $\Sigma n_i = 500$, so $S(\overline{D}_V) = s/\sqrt{500}$

Cols 2 & 3 Derived from the data for number of particles per unit volume from Schwartz–Saltykov (S-S) Analysis and Woodhead (W) analysis in Table 5.3 using eqn (A4) to obtain $\overline{D}_V$ and eqn (A7) to obtain its confidence limits.

spheres of discrete sizes $j\Delta$, and means that if a histogram is plotted, it should be shifted by $+\frac{1}{2}\Delta$ to correspond to the true volume size distribution. Conversely, in Fig. 5.4 the histogram plotted from the results of the Woodhead analysis corresponds closely with the true histogram. Again, this is as expected from the derivation of the coefficients, which assumes spheres in the size range $(j-1) \Delta$ to $j\Delta$, and has the advantage that it avoids any possibility of confusion.

In order to derive the mean size in the volume $(\overline{D}_V)$ and its confidence limits, the shift in the histogram for the Schwartz–Saltykov analysis must be recalled. Then, the results in Table 5.4 are obtained from Table 5.3 and eqns A4 and A7 in the Appendix.

From the calculations in Table 5.4, the Woodhead analysis gives $\overline{D}_V$ in agreement with the true value, and the Schwartz-Saltykov value is within the confidence limits.

Comments

1. If the results from the Schwartz–Saltykov analysis are incorrectly analysed as the histogram in Fig. 5.3, a value of $\overline{D}_V = 27.8$ µm would be obtained, i.e. half a size interval too low, which is a serious systematic error.

2.   The Schwartz–Saltykov analysis is the standard one used, because the Woodhead coefficients, Table 5.2, are from unpublished work. However, the Woodhead analysis has the advantage of accuracy, and analyses histograms in the usual way.

3.   From Fig. 5.1, the skewed nature of the area distribution, arising from the fact that large spheres contribute to all the smaller size groups of circles, is clear. This leads to a value of mean circle diameter in area, $\bar{d}_A = 26.1$ μm, indicating the importance of correct analysis to obtain the value of mean sphere diameter in volume, $\bar{D}_V = 30.0$ μm.

4.   Experimentally, many particle size distributions in the volume are skewed to small sizes, rather than approximating to normal distributions.

5.3 ANALYSIS OF EXTRACTION REPLICAS

In preparing an extraction replica it is important that a mechanically polished surface is first etched sufficiently deeply to remove all the sectioned particles, as shown in Fig. 5.5(a). The carbon is then deposited to form the replica, Fig. 5.5(b), which is released by the second etch and is floated off. Ideally, it will contain all the particles that impinged on the first etched surface, and no extraneous particles released by the first etch and not

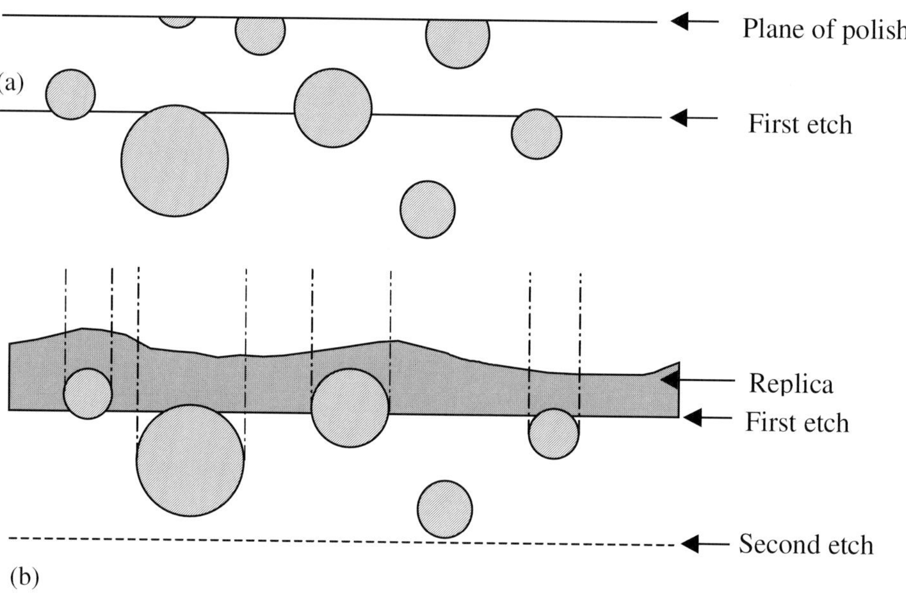

Fig. 5.5 Preparation of an extraction replica.

washed off the surface, or released by the second etch and not rinsed off the replica. In the analysis of replicas it is assumed that this ideal situation is achieved. In practice it requires skill and experience, and with normal replicas there is no contrast from the matrix, so the quality of the replica cannot be judged.

A modification of this usual procedure is to include shadowing of the surface after the first etch by sputtering gold at an acute angle to produce a shadowed extraction replica (Mukherjee, et al. (1968)), as shown in Fig. 5.6. A certain fraction (depending on the shadowing angle) of the particles now cast shadows. The fraction can be measured experimentally and compared with that expected for an ideal replica, to show that extraneous particles from the second etch, which cannot cast shadows, are not present. This is true for Fig. 5.6, from which the matrix contrast also shows that only two particles should have been present which were not extracted. From the images (arrowed) it is still possible to estimate a size. Importantly, the shadowing also shows that the matrix is not flat, but is ridged on the scale of the particle size so that the true surface area is considerably larger than the projected area of the micrograph.

Measurements of particle sizes from micrographs of extraction replicas can be made by the methods outlined in Section 5.1. For measurement by hand, the use of shadowed replicas has advantages indicated above, but for automatic measurement it is preferable to have no contrast in the background. In all cases, after a deep first etch, the observed images are from the true particle diameters in the volume, Fig. 5.5. No corrections of the sort required for plane sections are needed, but the measured numbers per unit area, $N_{A(j)}$, must be corrected to the numbers per unit volume, $N_{V(j)}$, by applying eqn (5.1). Thus for each size group in a histogram of size interval Δ.

$$N_{V(j)} = \frac{N_{A(j)}}{(j - \tfrac{1}{2})\Delta} \tag{5.6}$$

EXAMPLE 5.3.1 – PARTICLE SIZE FROM EXTRACTION REPLICAS

The micrograph in Fig. 5.6 was taken from a 0.2% plain carbon steel which had been quenched and tempered for one hour at 700°C and then cold worked and further tempered for 4 hours at 700°C in order to study carbide particle coarsening in the recrystallised ferrite matrix (Mukherjee et al. (1968)). From an enlarged version of this micrograph, of final magnification × 5700, the numbers of particles shown in Column 3 of Table 5.5 were obtained in each size group of 1 mm size interval from an area 182 × 201 mm.

From the magnification, the size interval

$$\Delta = \frac{1}{5700} = 0.1754 \mu m \text{ and the (projected) area measured}$$

$$A = \frac{182 \times 201}{(5700)^2} = 1.126 \times 10^{-3} \, mm^2$$

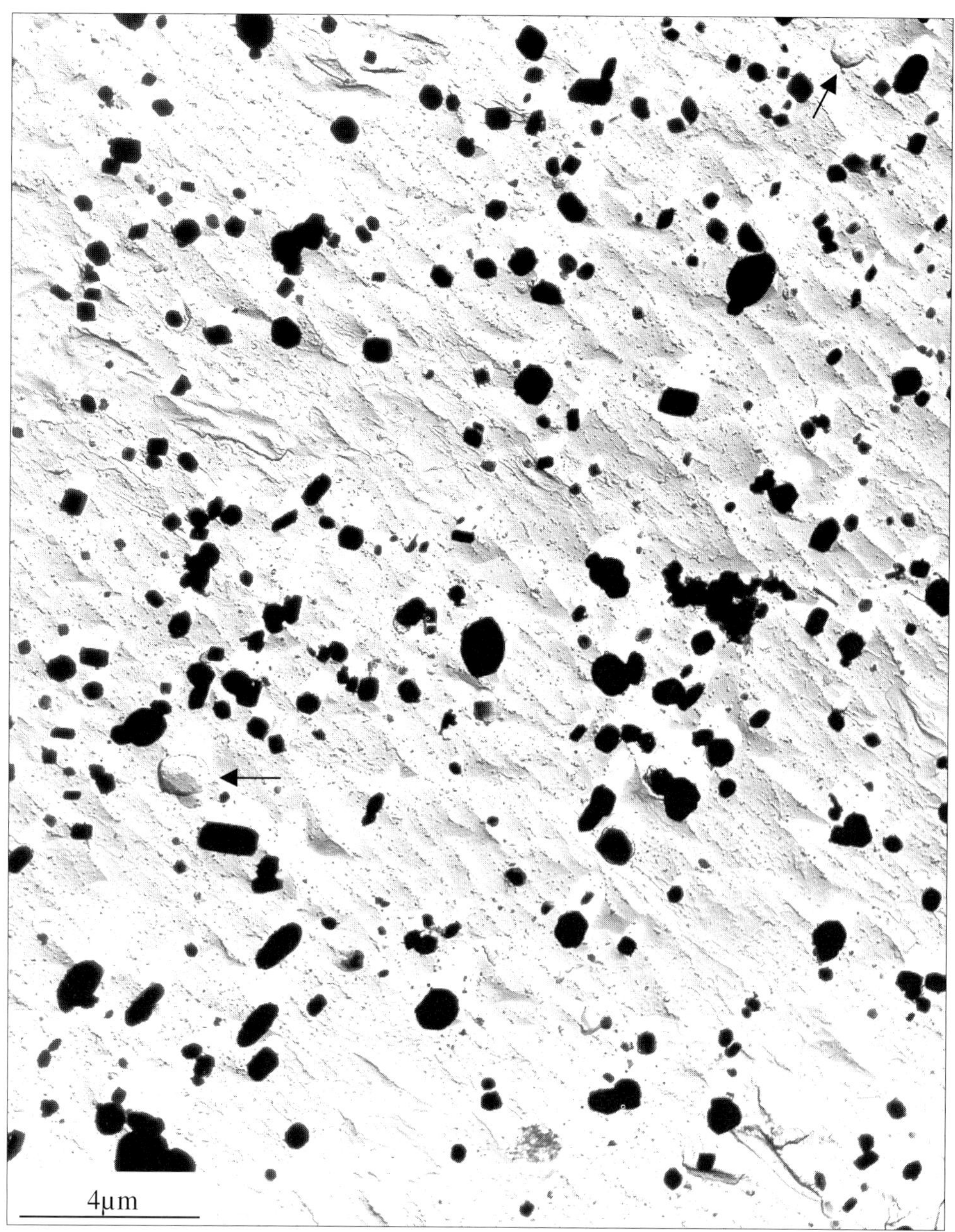

Fig. 5.6 Shadowed extraction replica of quenched and tempered 0.2% C steel.

Table 5.5 Analysis of carbide particle sizes from an extraction replica.

(1) Group No. j	(2) Upper Diameter (μm)	(3) Particle Nos. $n(j)$	(4) $N_A(j)$ (mm^{-2})	(5) $F_A(j)$ (%)	(6) $N_V(j)$ (mm^{-3})	(7) $F_V(j)$ (%)
1	0.175	55	48848	10.6	5.570×10^8	33.0
2	0.351	204	181181	39.1	6.886×10^8	40.8
3	0.526	147	130557	28.2	2.977×10^8	17.7
4	0.702	61	54177	11.7	0.883×10^8	5.2
5	0.877	34	30197	6.5	0.383×10^8	2.3
6	1.053	12	10658	2.3	0.111×10^8	0.7
7	1.228	5	4441	1.0	0.039×10^8	0.2
8	1.404	2	1776	0.4	0.014×10^8	0.1
9	1.579	1	888	0.2	0.006×10^8	0.0
		521	462723	100.0	16.869×10^8	100.0

Col 2 Upper circle diameters on replica for each size group.
Col 3 Experimental data for numbers in each size group.
Col 4 Numbers of particles per unit area derived from Col 3 by dividing by the projected area of replica.
Col 5 Frequencies in each size group from Col 4 divided by the total number of particles per unit area.
Col 6 Numbers of particles per unit volume from Col 4 and Equation (5.6).
Col 7 Frequencies in each size group from Col 6 divided by the total number of particles per unit volume.

From the particle numbers in Column 3,

$$N_A(j) = \frac{n(j)}{A}$$

and $\quad F_A(j) = \left(\frac{N_A(j)}{\Sigma N_A(j)} \right) \times 100$

shown in Columns 4 and 5 were obtained. By applying eqn (5.6) the values of $N_V(j)$ in Column 6 are found, and hence the frequencies in the volume, $F_V(j)$, in Column 7.

The frequencies have been calculated so that the area and volume size distributions can be compared, as shown in Fig. 5.7. It is clear that the volume distribution is more skewed to smaller sizes than the area distribution, and it approximates more closely to a log. normal distribution than to a normal distribution.

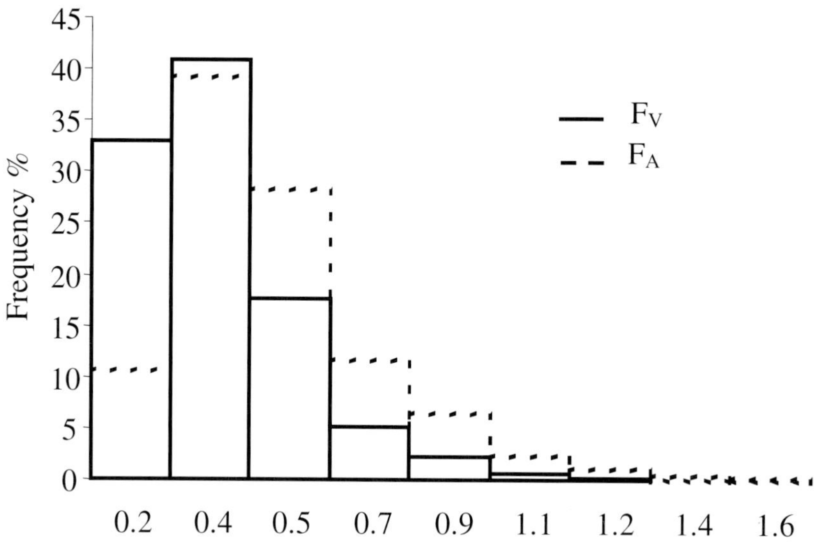

Fig. 5.7 Particle size distributions calculated for area and volume in Table 5.4.

Table 5.6: Statistical parameters calculated from the data in Table 5.5.

$\Sigma N_v(j) \cdot (j - \tfrac{1}{2}) \Delta$	4.6289×10^{5} mm^{-2}
$\Sigma N_v(j)$	1.6869×10^{9} mm^{-3}
$\overline{D}_v$	2.744×10^{-4} mm
$\Sigma N_v(j) \cdot ((j - \tfrac{1}{2}) \Delta)^2$	1.8549×10^{2} mm^2
$\overline{D}_v^2\ \Sigma N_v(j)$	1.2702×10^{2} mm^2
s	1.862×10^{-4} mm
$S(\overline{D}_v)$	8.156×10^{-6} mm

The mean and confidence limits for $\overline{D}_V$ are determined from the data in Column 6 of Table 5.5 and eqns (A4) and (A7) in the Appendix to give the values shown in Table 5.6.

Thus $\overline{D}_V = 2.74 \pm 0.016 \times 10^{-4}$ mm.

The same answer is obtained if the original numbers from the micrograph are used. From eqns (5.3) and (5.6) it can be shown by simple manipulation that

$$\Sigma N_V(j) \cdot (j - \tfrac{1}{2}) \, \Delta = \Sigma N_A(j) = \frac{M^2}{A} \, \Sigma n_j \tag{5.7}$$

$$\Sigma N_V(j) = \frac{M^2}{A\Delta} \cdot \Sigma \frac{n_j}{(j - \tfrac{1}{2})} \tag{5.8}$$

$$\text{Hence } \overline{D}_v = \frac{N_A}{N_v} = \frac{\Delta \cdot \Sigma nj}{\Sigma \dfrac{nj}{(j - \tfrac{1}{2})}} \tag{5.9}$$

and from eqn (A.7)

$$s^2 \approx \Delta^2 \left\{ \frac{\Sigma n_j (j - \tfrac{1}{2})}{\Sigma \dfrac{n_j}{(j - \tfrac{1}{2})}} - \left(\frac{\Sigma n_j}{\Sigma \dfrac{n_j}{(j - \tfrac{1}{2})}} \right)^2 \right\} \tag{5.10}$$

From Columns 1 and 3 in Table 5.5,

$$\Sigma n_j = 521$$

$$\Sigma \frac{n_j}{(j - \tfrac{1}{2})} = 333.1$$

$$\Sigma n_j (j - \tfrac{1}{2}) = 1189.5$$

and substitution into eqns (5.9) and (5.10) leads to the same values for $\overline{D}_V$ and its confidence limits as before.

Comments

1. The value of A does not appear in eqns (5.9) and (5.10), so that the use of projected area of the micrograph has no effect on the calculated values of mean size, or its confidence limits.

2. In the measured micrograph, the shadowing shows that the observed particles are as expected for the surface after the first etch. It is notable, however, that if particles are not extracted, i.e. they are pulled off the replica when it is floated after the second etch; they tend to be towards the top end of the size range. Conversely, extra particles released by the second etch, i.e. those contained within the depth of the second etch, and not rinsed off the replica tend to be towards the bottom end of the size range. Good replica preparation technique is therefore essential to avoid systematic errors.

3. If only the mean diameter of particles in the volume ($\bar{D}_V$) and its confidence limits are of interest, Ashby and Ebeling (1966) showed that these can be calculated directly from the mean particle size observed on the replica and its standard deviation, without the need to calculate the size distribution in the volume.

4. The apparent total number of particles per unit volume (Column 6 of Table 5.5) of 1.69×10^9 mm^{-3} may be compared with the value of 1.3×10^9 mm^{-3} obtained from analysis of thin foils of the same material given in the next example. This indicates an excess of about 30% from the replica, which would lead to at least as large an error in volume fraction if this were (erroneously) estimated from measurements on the replica.

5.4 ANALYSIS OF THIN FOILS

When particles are mainly less than 50 nm in diameter, measurements are best made by transmission electron microscopy of thin foils. In metallic systems, small second phase particles are usually more chemically inert than the matrix, so electrolytic dissolution of the matrix to produce the thin foil leaves the particles unaffected. Some particles will therefore stick out of the foil surfaces, as illustrated in Fig. 5.8.

In this case, which is the only one that will be considered, it is assumed that all particles with their centres inside the foil and up to a particle radius outside the foil will be observed. It is also assumed that a particle impinging on the surface does not perturb the local dissolution rate of the matrix, so the foil surfaces remain flat, as illustrated in Fig. 5.8. It is important that the thin foil preparation procedure ensures that particles released during matrix dissolution have been rinsed away.

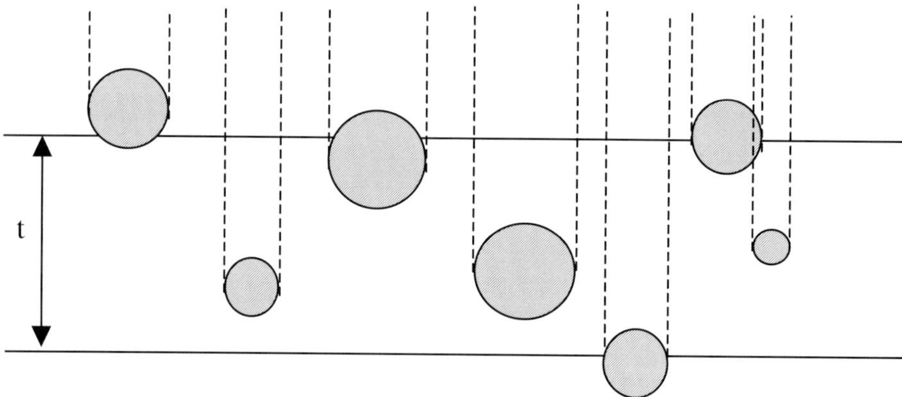

Fig. 5.8 Schematic illustration of a cross-section of a thin foil containing inert second phase particles.

Measurements of particle sizes from micrographs of thin foils can be made by the methods outlined in Section 5.1. If the particles are incoherent with the matrix this is usually done from bright field images, but for coherent particles the coherency strain contrast may interfere with size measurements. This difficulty can be avoided by using dark field images obtained from the coherent particles.

Particle images are then observed at their true volume diameter, except occasionally when a small particle may be partially obscured by a larger particle at a different level in the foil, as shown for the small particle to the right of Fig. 5.8. The number of particles observed in any size group, $N_{A(j)}$, will then be

$$N_{A(j)} = N_{V(j)}\left(t + D_{(j)}\right) - M_A \tag{5.11}$$

where $N_{V(j)}$ is the number of particles per unit volume in size group j, D_j is the mean diameter in the size group, $(j - \frac{1}{2})\Delta$, and t is the foil thickness.

The local foil thickness must therefore be measured using one of the methods described in textbooks on electron microscopy, e.g. (Loretto (1994)). D_j appears in the equation because the effective thickness from which particles are observed arises from particles with centres up to a radius above the top surface and a radius below the bottom surface in addition to those with centres inside the foil.

M_A is a correction factor for the number of images lost by overlap. This correction has been considered in detail by Hilliard (1962). The treatment is complex and the correction depends on the actual particle distribution and on the extent to which overlapping images can be resolved. However, it may be ignored for spherical particles, if an error of up to 5% in N_V is acceptable, when

$$\frac{V_V t}{\overline{D}_V} < 0.04 \tag{5.12}$$

This condition should be checked when measurements have been made, but it covers many practical situations when the volume fraction of particles, V_V, is low, if $\overline{D}_V$ is of the same order of size as t. If $\overline{D}_V < t$, serious errors in interpretation of the observed size distribution could arise even for low volume fractions. In the example in this section, as shown later, the condition in eqn (5.14) is satisfied.

EXAMPLE 5.4.1 – PARTICLE SIZE FROM THIN FOIL

The micrograph in Fig. 5.9 is from a thin foil of the same 0.2% plain carbon steel given the same thermomechanical treatment as for the micrograph shown in Fig. 5.6, i.e. the steel was quenched and tempered for one hour at 700°C and then cold worked and further tempered for four hours at 700°C. In general, the particle sizes can be measured unambiguously, but at the bottom of the Figure there is an image which is probably best interpreted as arising from partially overlapping images from three particles, as indicated by the drawn-in white boundary lines. With this assumption, the number of particles

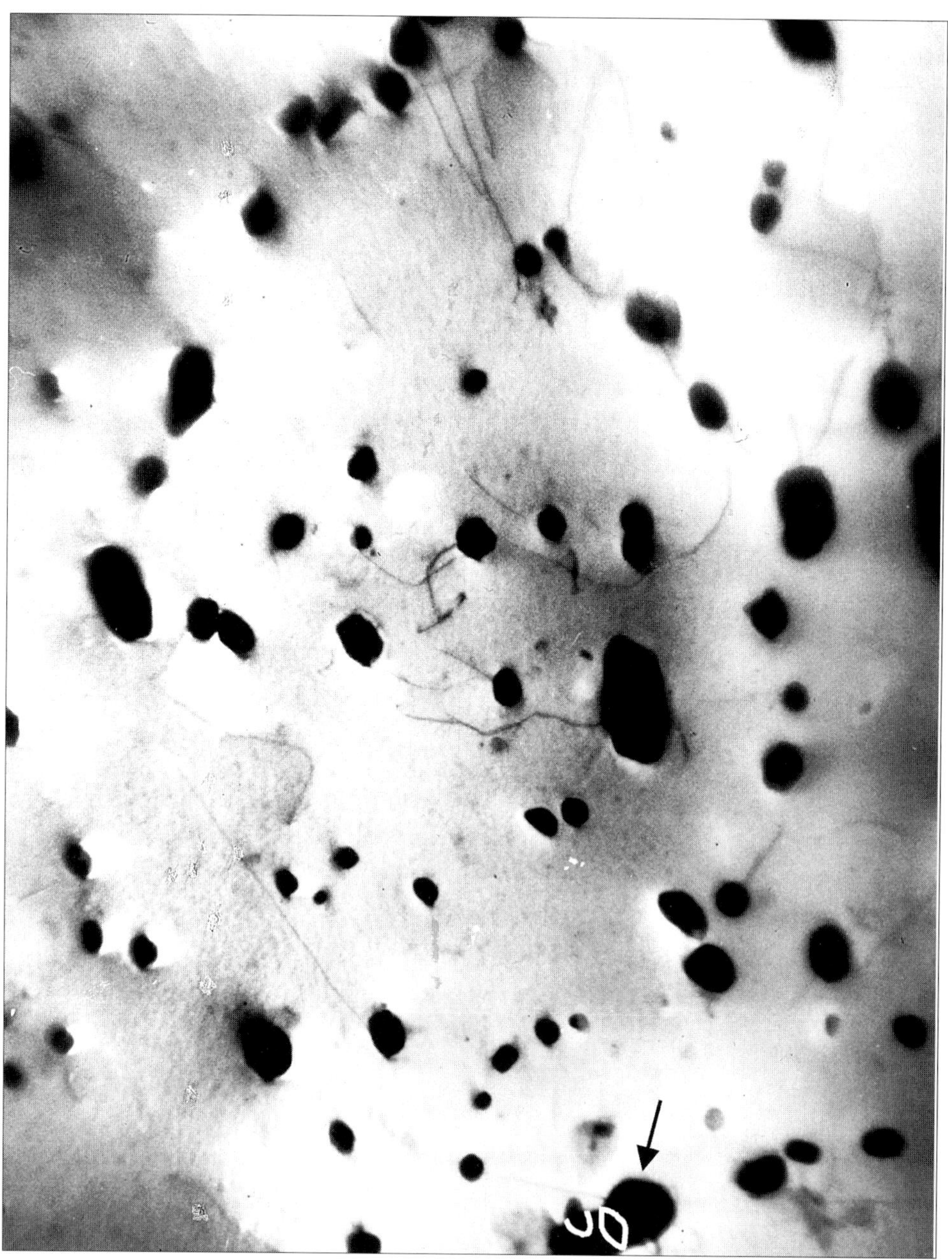

Fig. 5.9 Micrograph taken from a thin foil of quenched and tempered 0.2% C steel.

Table 5.7 Analysis of carbide particle sizes from the thin foil electron micrograph in figure 5.9.

(1)	(2)	(3)	(4)	(5)	(6)	(7)
Group No (j)	Upper Diameter (μm)	Particle Nos. $n(j)$	$N_{A(j)}$ (mm^{-2})	$F_{A(j)}$ (%)	$N_{V(j)}$ (mm^{-3})	$F_{V(j)}$ (%)
1	0.108	5	39530	6.2	1.556×10^8	11.7
2	0.216	11	86966	13.6	2.401	18.1
3	0.324	33	260897	40.6	5.548	41.7
4	0.432	16	126496	19.8	2.187	16.4
5	0.541	6	47436	7.4	0.691	5.2
6	0.649	6	47436	7.4	0.597	4.5
7	0.757	2	15812	2.5	0.175	1.3
8	0.865	0	0	0	0	
9	0.973	2	15812	2.5	0.141	1.1
Total		**81**	**6.404×10^5**	**100.0**	**13.30×10^8**	**100.0**

Col 2 Upper circle diameters in the thin foil for each size group,
Col 3 Experimental data for numbers in each size group,
Col 4 Numbers of particles per unit area derived from Col 3 by dividing by the observed area of the foil,
Col 5 Frequencies in each size group from Col 4 divided by the total number of particles per unit area,
Col 6 Numbers of particles per unit volume from Col 4 and Equation (5.13),
Col 7 Frequencies in each size group from Col 6 divided by the total number of particles per unit volume.

measured in 2 mm size groups on the original micrograph at a magnification of ×18,500, which gave an area of 234 × 185 mm is shown in Column 3 of Table 5.7.

From the magnification, the size interval

$$\Delta = \frac{2}{18500} = 1.081 \times 10^{-4} \text{ mm}$$

$$= 0.1081 \text{ }\mu\text{m}$$

and $\quad A = \dfrac{234 \times 185}{(18500)^2} = 1.2649 \times 10^{-4} \text{ mm}^2$

From the particle numbers in Column (3) the number per unit area in each size group

$$N_{A(j)} = \frac{n(j)}{A}$$

giving the values shown in Column 4.

The values of numbers of particles per unit volume $N_{V(j)}$, in each size group shown in Column 6 are then simply calculated from the measured foil thickness of 0.20 µm $(2 \times 10^{-4}$ mm$)$ and eqn (5.11) as

$$N_{V(j)} = \frac{N_{A(j)}}{\left(t+(j-\frac{1}{2})\Delta\right)} \tag{5.13}$$

The frequencies of occurrence of particles per unit area and per unit volume are shown in Column 5 and 7, respectively. It is clear that the frequency of occurrence of small particles in the volume is greater than observed in the area and that the situation is reversed for large particles.

The mean and confidence limits of $\overline{D}_V$ are determined from the data in Column 6 and eqns (A4) and (A7) in the Appendix to give the values shown in Table 5.8:

From this single micrograph

$$\overline{D}_v = 2.82 \pm 0.34 \times 10^{-4} \text{ mm}$$

$$= 0.282 \pm 0.034 \mu\text{m}$$

The volume fraction of particles can be estimated directly because the measured equivalent circle diameters correspond to sphere diameters in the volume. Hence:

$$V_v = \frac{\pi}{6}\Delta^3 \, \Sigma N_{v(j)}(j-\frac{1}{2})^3 \tag{5.14}$$

From the values of $N_v(j)$ in Column 6 of Table 5.6 and $\Delta^3 = 1.263 \times 10^{-12}$ mm^3,

$$V_V = 0.0321$$

Because 81 particles is a small sample from a very small volume, measurements were made on 2 additional areas on the same foil and 3 areas in a different foil. These were analysed as above, to obtain the values of $\overline{D}_V$ and V_V given in Table 5.9.

The overall mean value of particle size

Table 5.8 Statistical parameters calculated from the data in Table 5.7.

$\Sigma N_{V(j)} \cdot (j-\frac{1}{2})\Delta$	3.74×10^{5} mm^{-2}
$\Sigma N_{V(j)}$	1.330×10^{9} mm^{-3}
$\overline{D}_V$	2.82×10^{-4} mm
$\Sigma N_V(j) \cdot ((j-\frac{1}{2})\Delta)^2$	1.366×10^{2} mm^{2}
$\overline{D}_V^{\,2} \Sigma N_V(j)$	1.058×10^{2} mm^{2}
s	1.532×10^{-4} mm
$S(\overline{D}_v) = \dfrac{s}{\sqrt{81}}$	1.70×10^{-5} mm

Table 5.9 Particle measurements on six regions of thin foil.

(1)	(2)	(3)	(4)	(5)
Region	$n(z)$	$\overline{D}_V(z)$ μm	$N_v(z)$ mm^{-3}	$V_v(z)$
1	81	0.282 ± 0.034	1.33 × 10^9	0.0321
2	72	0.276 ± 0.036	1.06 × 10^9	0.0223
3	104	0.261 ± 0.030	1.47 × 10^9	0.0271
4	86	0.264 ± 0.033	1.38 × 10^9	0.0264
5	92	0.278 ± 0.032	1.46 × 10^9	0.0314
6	80	0.279 ± 0.034	1.15 × 10^9	0.0250

Col 2 Numbers of particles measured in each region of the foil,
Col 3 Mean particle diameter in the volume calculated for each region of the foil,
Col 4 Numbers of particles per unit volume calculated for each region of the foil,
Col 5 Volume fraction of particles calculated for each region of the foil.

$$\overline{D}_V = \frac{\Sigma n(z)\overline{D}_v(z)}{\Sigma n(z)} \tag{5.15}$$

substituting the values in Columns 2 and 3 gives:

$\overline{D}_V = 0.2727$ μm

Considering that each region gives an independent measure, and applying the statistical procedures in the Appendix to the variable $n(z)\,\overline{D}_V(z)/\Sigma\,n(j)$ (i.e. weighting each region according to the number of particles) gives a standard deviation

$s = 0.0300$ μm

Hence the standard error of the mean

$S(\overline{D}_V) = 0.0122$ μm

For 5 degrees of freedom $t_{95,5} = 2.571$ giving 95% confidence limits of ±0.031 μm. Hence,

$\overline{D}_V = 0.273 \pm 0.031$ μm

Because a single value of N_v and V_v is determined from each region of the foil, the measurements are not weighed to determine the means and confidence limits as

$\overline{N}_V = 1.31 \pm 0.18 \times 10^9$ mm^{-3}

and

$\overline{V}_V = 0.0274 \pm 0.0039$

Comments

1. From the results for region 1, the value of $V_v t/\bar{D}_V$ is $0.0321 \times 0.2/0.282 = 0.023$, which clearly satisfies the criterion of eqn (5.14). Thus the analysis carried out by assuming that overlapping of images can be neglected should be valid.

2. The mean particle size of 0.273 µm determined from the thin foils coincides closely with the value of 0.274 µm obtained from analysis of the extraction replica of the same material in the previous example.

3. The mean volume fraction of 0.0274 may be compared directly with the expected volume fraction of carbide in equilibrium at 700°C, because the long tempering time applied to the specimen means that equilibrium should have been achieved. Consider that in the 0.2%C plain carbon steel 0.02% C remains in solution at 700°C. The atomic weights of carbon and iron are 12 and 56, respectively, and the molar volumes of iron in ferrite and of iron carbide (Fe_3C) are 7.3 and 23.4 mm³, respectively.

In 100 g of steel, 0.18 g of carbon provides the Fe_3C particles.

$$\text{The volume of } Fe_3C = \frac{0.18 \times 23.4}{12} = 0.351 \text{cm}^3$$

$$\text{The weight of Fe in } Fe_3C = 3 \times 0.18\frac{56}{12} = 2.52 \text{g}$$

$$\text{The volume of Fe in the matrix} = \frac{98.8 - 2.52}{56} \times 7.3 = 12.551 \text{m}^3$$

The equilibrium volume fraction of Fe_3C is therefore

$$V_v = \frac{0.351}{(12.551 + 0.351)} = 0.0272$$

This value is in close agreement with the measured value.

5.5   INTERPARTICLE SPACING

In relating properties to microstructure, interparticle spacing rather than particle size is often the critical parameter. However, interparticle spacing may be defined in different ways, so care must be exercised in selecting the appropriate definition for a specific purpose. In all cases the spacings are not measured directly from micrographs, but are calculated from the particle parameters derived in the earlier examples in this chapter.

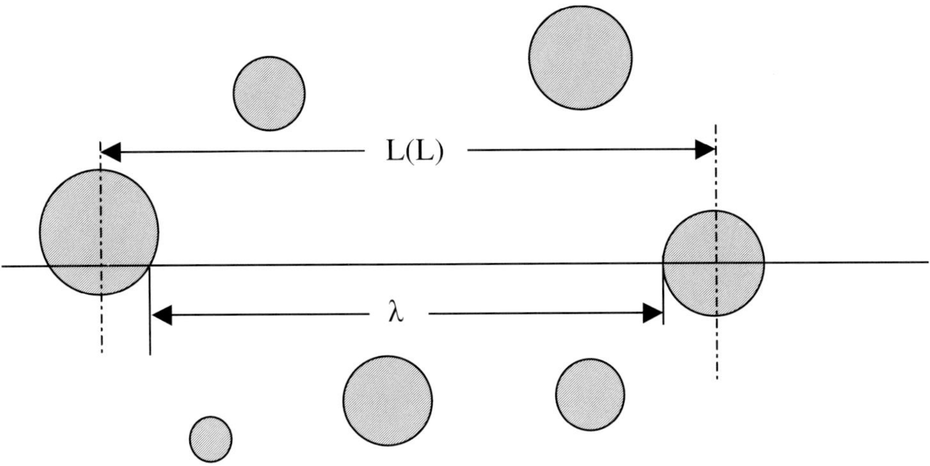

Fig. 5.10 Interparticle spacing along a random line drawn in the volume.

EXAMPLE 5.5.1 – SPACING ALONG A LINE

The mean centre to centre spacing between particles measured along a straight line in the volume, shown in Fig. 5.10, is simply the reciprocal of the number of particles per unit length, i.e.

$$\overline{L}(L) = \frac{1}{\overline{N}_L} \tag{5.16}$$

Alternatively the mean free distance λ between particles is given by

$$\lambda = \frac{(1 - V_V)}{\overline{N}_L} \tag{5.17}$$

where V_V is the volume fraction, which is equal to the line fraction of particles.

From the relationships for particle parameters derived by Fullman (1953):

$$\overline{N}_L \approx \frac{\pi}{4} \overline{N}_V \overline{D}_V^2 \tag{5.18}$$

The results of measurements of carbide particles in thin foils of a tempered steel, Example 5.4.1, gave

$$\overline{D}_V = 0.273 \ \mu m = 0.273 \times 10^{-3} \ mm$$

$$\overline{N}_V = 1.31 \times 10^9 \ mm^{-3}$$

$$\overline{V}_V = 0.0274$$

Hence, substitution into eqn (5.18) gives

$$\bar{N}_L = 76.7 \text{ mm}^{-1}$$

and substitution into eqns (5.16) and (5.17) gives

$$\bar{L}(L) = 0.0130 \text{ mm} = 13 \text{ μm}$$

and

$$\lambda = 0.0127 \text{ mm} = 13 \text{ μm}$$

Comments

1. Because of the low volume fraction of particles in this example, there is no significant difference between $\bar{L}(L)$ and λ.
2. The results from extraction replicas give a high apparent value of N_V so would lead to underestimation of interparticle spacing.

EXAMPLE 5.5.2 – SPACING IN A PLANE

From Fig. 5.10 it is apparent that the spacing between a particle and (the section) of another particle lying on the same plane but is any direction is smaller than the spacing along a specified line in the plane. The mean interparticle spacing on the plane is then

$$\bar{L}(A) = 0.5 \left(\frac{1}{N_A} \right)^{\frac{1}{2}} \tag{5.19}$$

For randomly distributed particles, where N_A is the number of particles (sections) per unit area,

$$N_A = N_V \bar{D}_V \tag{5.20}$$

From the results of measurements of carbide particles in thin foils of a tempered steel, Example 5.4.1,

$$\bar{D}_V = 0.273 \text{ μm} = 0.273 \times 10^{-3} \text{ mm}$$

$$\bar{N}_V = 1.31 \times 10^9 \text{ mm}^{-3}$$

Hence, from eqn (5.20).

$$\bar{N}_A = 3.58 \times 10^5 \text{ mm}^{-2}$$

and substituting into eqn (5.19) gives

$$\bar{L}(A) = 8.36 \times 10^{-4} \text{ mm} = 0.84 \text{ μm}$$

Comments

1. The very small value of $\bar{L}(A)$ compared with $\bar{L}(L)$ determined from the same measurements in Example 5.5.1 emphasises the critical nature of defining the

appropriate value of interparticle spacing for comparison with other published data, or with the theoretical effect of particles on properties.

2. Spacing of particles in a plane is particularly important in the interpretation of mechanical properties, because dislocations interact with particles that intersect their crystallographic slip planes. For comparison with theory, which assumes a uniform interparticle spacing, Kocks (1966) points out that for a random array of particles, the mean distance between a particle and its nearest two or three neighbours in the plane rather than with its nearest neighbour, should be applied, leading to a value of the constant in eqn (5.19) of 1.18 instead of 0.5.

EXAMPLE 5.5.3 – SPACING IN THE VOLUME

For consideration of diffusion fields and kinetic effects, the interparticle spacing in the volume, i.e. the distance in any direction from one particle centre to a neighbouring particle centre is the appropriate one. The mean value is then

$$\overline{L}(V) = 0.554 \left(\frac{1}{N_V} \right)^{\frac{1}{3}} \tag{5.21}$$

From the results of measurements of carbide particles in thin foils of a tempered steel, Example 5.4.1,

$$\overline{N}_V = 1.31 \times 10^9 \, \text{mm}^{-3}$$

Hence from eqn (5.21)

$$\overline{L}(V) = 5.10 \times 10^{-4} \, \text{mm} = 0.51 \, \mu\text{m}$$

Comments

1. Considering the confidence limits of the measurements of $\overline{N}_V$ of $\pm 0.18 \times 10^9 \, \text{mm}^{-3}$ leads to values of $L(V)$ in the range of 0.485 to 0.532, i.e. $\overline{L}(V) = 0.51 \pm 0.02 \, \mu\text{m}$. This is equivalent to relative confidence limits of one third of the relative confidence limits of $\overline{N}_V$, as expected from eqn (5.21).

2. The value of $\overline{L}(V)$ is again significantly smaller than the value of $\overline{L}(A)$ obtained in Example 5.5.2, but in this case is of the same order of magnitude.

6. Dislocation Structure from Thin Foil Electron Micrographs

During deformation, work hardening occurs because the dislocation density increases with increasing strain, and the flow stress (σ) increases with total dislocation density, i.e. length of dislocation lines per unit volume, (ρ) as (Dieter (1986))

$$\sigma = \sigma_i + \alpha\mu\mathbf{b}\rho^{\frac{1}{2}} \tag{6.1}$$

where σ_i is the 'friction stress', μ is the shear modulus, $\mathbf{b}$ is the Burgers vector and α is a constant.

As the strain increases the dislocations become arranged in 'cells' with boundary regions of high density and interiors of low density. At elevated temperatures, recovery enables the cell boundaries to rearrange into near two dimensional 'sub-grain' boundaries of mean spacing, δ, with misorientations, θ, across them. The flow stress can then be expressed as

$$\sigma = \sigma_i + \alpha_1\mu\mathbf{b}\rho_i^{\frac{1}{2}} + \alpha_2\mu\mathbf{b}\delta^{-1} \tag{6.2}$$

where ρ_i is the dislocation density inside sub-grains and α_1 and α_2 are constants.

Following deformation, under conditions where recrystallisation is suppressed, a small proportion of the energy of deformation is stored by the microstructure in the form of dislocations and dislocation structures, for example sub-grain boundaries etc. and a measure of stored energy per unit volume in a deformed material can be derived as (Sellars and Zhu (1999))

$$E \approx \frac{\mu\mathbf{b}^2}{10}\left[\rho_i\left(1 - \ln\left(10\mathbf{b}\rho_i^{1/2}\right)\right) + \frac{2\theta}{\mathbf{b}\delta}\left(1 - \ln\frac{\theta}{\theta_c}\right)\right] \tag{6.3}$$

where θ_c is the critical sub-grain misorientation when the sub-grain boundary can be considered as equivalent to a grain boundary; typically 15°. Each of these dislocation parameters can be measured directly by careful thin foil analysis, as described in the following sections.

6.1 DISLOCATION DENSITY

There are two methods for measuring the dislocation density; end counting and the intercept method. The method used depends on the dislocation density, type of material, the skill of the operator and the accuracy required. For both of the methods described below, an electron micrograph of the area of interest is taken, usually with multi-beam conditions, to allow the maximum fraction of dislocations to be in contrast.

EXAMPLE 6.1.1 – DISLOCATION DENSITY BY END COUNTING

Method

If it is possible to image individual dislocations, for example Fig. 6.1, then a simple measure of the dislocation density can be determined by counting the 'ends' of the dislocations. These 'ends' are produced by the intersection of the dislocation by the top and bottom surfaces of the foil.

The dislocation density $\rho = L_V$, the line length per unit volume, is then given by

$$L_V = 2\overline{P}_A \tag{6.3}$$

where $\overline{P}_A$ is the number of points per unit area of intersection of dislocations with a plane, assuming random orientation of the dislocations with respect to the plane.

If n dislocation ends are counted in an area of foil A, then the total area intersected is $2A$ because the top and bottom surfaces are both of area A. Thus,

$$\overline{P}_A = \frac{n}{2A} \tag{6.4}$$

Hence, from eqn 6.3

$$\rho_i = \frac{n}{A} \tag{6.5}$$

This equation requires the area of the micrograph to be measured, corrected for magnification, M, and so for measurements made directly from the micrograph, eqn (6.5) can be written as

$$\rho_i = \frac{nM^2}{l_1 l_2} \tag{6.6}$$

where l_1 and l_2 are the lengths of the two sides of the micrograph/negative in metres.

Results

For the sample in Fig. 6.1 the number of ends visible is 203 in an area 0.090×0.071 m at a magnification of $105\,000\times$. This gives a dislocation density of:

$$\rho_i = \frac{203 \times 105\,000^2}{0.090 \times 0.071}$$

$$\rho_i = 3.55 \times 10^{14}\,\text{m}^{-2}.$$

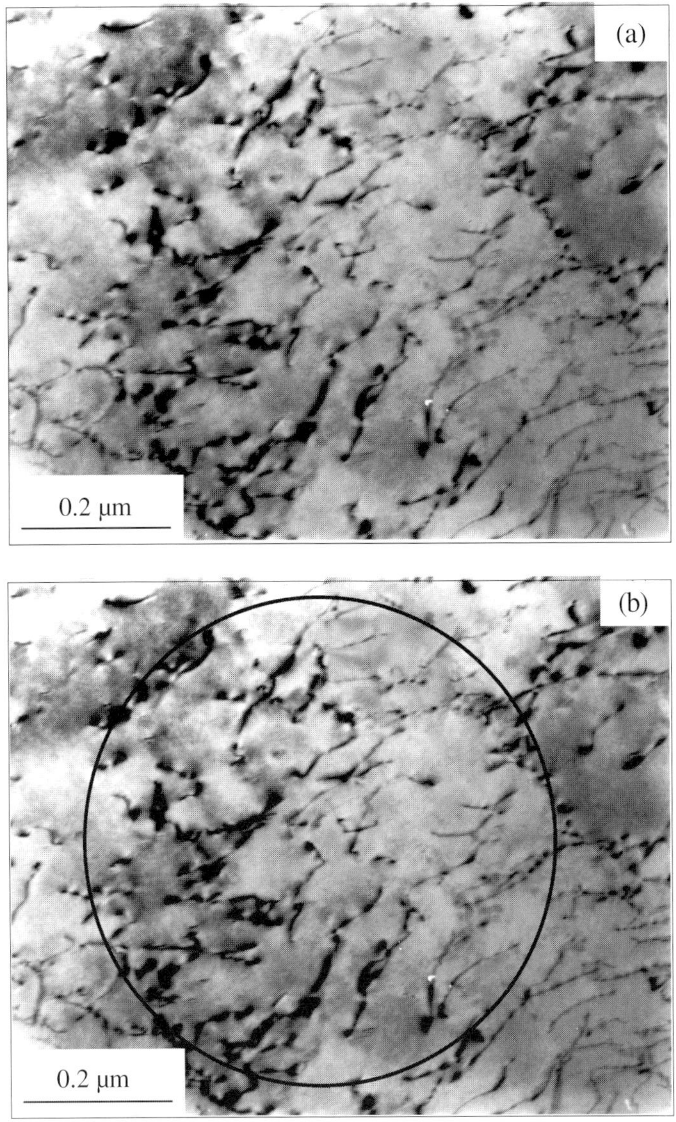

Fig. 6.1 Dislocations in hot rolled stainless steel (a) micrograph for counting ends and (b) circle used for intercept method.

Table 6.1 Dislocation densities measured by 'end counting'.

(1) Electron Micrograph	(2) Number of Dislocation Ends, n	(3) Dislocation Density (m^{-2})
1	203	3.55×10^{14}
2	177	3.10×10^{14}
3	222	3.89×10^{14}
4	213	3.73×10^{14}
5	230	4.03×10^{14}
6	201	3.52×10^{14}
7	186	3.26×10^{14}
8	170	2.98×10^{14}
9	197	3.45×10^{14}
10	159	2.78×10^{14}

Col 2 Experimental measurements.
Col 3 derived from Col 2 and eqn 6.6.

Nine further electron micrographs were taken in different sub-grains in the same foil and in different foils, and the results are shown in Table 6.1.

The mean and standard deviation for these measurements are:

$$\rho_i = 3.429 \times 10^{14}\,m^{-2}$$

$$s = 0.287 \times 10^{14}\,m^{-2}$$

Which gives $t_{95,\,9}$ confidence limits of

$$\rho_i = 3.5 \pm 0.3 \times 10^{14}\,m^{-2}$$

Comments

1.   The dislocation structure in the foils used in this example is surprisingly uniform, nevertheless it can be seen from Column 3 of Table 6.1 that dislocation densities measured from individual foils may vary significantly. This is considered further in the comments on Example 6.1.2.

Example 6.1.2 – Dislocation Density by Intercept Method

The advantage of this technique is that it can be used in thicker parts of the foils where the dislocation ends cannot be easily distinguished. The technique does, however, require the thickness of the foil to be measured. The method for measuring the foil thickness will not be discussed here, but can be found in electron microscopy textbooks, e.g. Loretto (1994).

Method

On an electron micrograph of the area of interest, a circle of known circumference is randomly placed as shown in Fig. 6.1b. The number of intercepts of dislocations with the circle is then counted and the circle is moved or placed on the next micrograph. The dislocation density is then given by

$$\rho_i = \frac{2nM}{Lt} \tag{6.7}$$

where n is the number of intercepts with the circle, M is the magnification, L is the circumference of the circle in the micrograph and t is the foil thicknes, i.e. Lt/M is the area of the cylinder through the foil intersected by the dislocations.

Results

In Fig. 6.1b, a circle of circumference 200 mm is placed over the micrograph, of magnification 105 000× and the number of intercepts with the circle is 29. The local foil thickness was determined as 120 nm, which gives a dislocation density of

$$\rho_i = \frac{2 \times 29 \times 105000}{0.2 \times 1.2 \times 10^{-7}}$$

$\rho_i = 2.54 \times 10^{14}\,\mathrm{m}^{-2}$.

As for the previous example nine further measurements were made on micrographs taken in different sub-grains in the same and different foils, and the results are shown in Table 6.2.

The mean and standard deviation for these measurements are

$\rho_i = 2.94 \times 10^{14}\,\mathrm{m}^{-2}$

$s = 0.588 \times 10^{14}\,\mathrm{m}^{-2}$

Which gives a $t_{95,9}$ confidence interval of

$\rho_i = 2.9 \pm 0.42 \times 10^{14}\,\mathrm{m}^{-2}$

Comments

1. It can be seen that the results of the two methods of measuring ρ_i give similar values. The differences arise from the area covered using each method. With the end counting method a large area of the foil is used for the observations. With the

Table 6.2 Dislocation densities measured by the intercept method.

(1)	(2)	(3)	(4)
Electron Micrograph	Number of Intersections, n	Foil Thickness, t (nm)	Dislocation Density (m^{-2})
1	29	120	2.54×10^{14}
2	17	90	2.02×10^{14}
3	43	156	2.89×10^{14}
4	38	113	3.56×10^{14}
5	27	95	2.96×10^{14}
6	33	102	3.36×10^{14}
7	37	134	2.90×10^{14}
8	46	126	3.83×10^{14}
9	55	178	3.24×10^{14}
10	30	147	2.13×10^{14}

Col 2 Experimental measurements of number of intersections with a circle,
Col 3 Experimental measurements of local foil thickness,
Col 4 Derived from Col 2, Col 3 and eqn (6.7).

intercept method, however, only the dislocations intersecting the smaller area of the cylinder projected as the circle are considered. This method is needed when measuring the dislocation density in thicker parts of the foil, where it is not possible to identify individual dislocation ends.

2. Other errors that arise when using this technique occur in the measurement of the foil thickness. Due to the profile of foil edges, the thickness is not constant over the area of the electron micrograph. In the present case a measure of the foil thickness at the centre of the electron micrograph was used. A more precise thickness profile could be obtained by measuring the thickness of the foil at the four corners of the micrograph.

3. The results obtained in the example are very uniform, as the dislocation density observed in the stainless steel was homogeneous. Other metals, for example aluminium, may not exhibit such a uniform dislocation structure. Care must be taken to ensure that micrographs provide a good statistical sample, see Chapter 1, Section 1.2.4.

4. Regardless of the technique used to measure the density, electron micrographs should be taken from a number of sub-grains and in several different foils to obtain a statistically meaningful average. This is necessary because of variations in dislocation density between sub-grains and variations between grains of different orientation.

5. Care must be taken to avoid the possibility of dislocations being introduced into the foils during foil preparation, e.g. by bending the foil.

6.2 SUB-BOUNDARY MISORIENTATION

There are two techniques for the measurement of sub-boundary misorientation. These are from a thin foil in the TEM and from a surface by Electron Back Scattered Diffraction (EBSD). EBSD is currently restricted in its usage, however, because of its limited resolution both in the size of features imageable (> 70 nm) and, more importantly, its angular resolution (1°), and will not therefore be discussed further here.

There are a number of methods of measuring the misorientation in thin foils. The method used depends mainly on the degree of accuracy required, the equipment available and the skill of the operator. Two methods will be discussed here, both of which use Convergent Beam Electron Diffraction (CBED).

EXAMPLE 6.2.1 – KIKUCHI PATTERN DISPLACEMENT

Method

A typical sub-grain structure in aluminium is shown in Fig. 6.2. This micrograph is taken at a multi beam condition, as for measuring the dislocation density described in Section 6.1. Initially a reference sub-grain is chosen in the area of interest, for example sub-grain A in Fig. 6.2, which is tilted to a known zone axis using CBED, for example in fcc metals the zone axis $B = [100], [111]$ etc. (Loretto (1994)) and an image taken. The beam is then moved to a neighbouring sub-grain and, without further tilting, an image of the diffraction pattern is taken. The procedure is repeated for all of the neighbouring sub-grains. A new reference sub-grain is then selected and tilted to a known zone axis and the procedure repeated.

A schematic of the Kikuchi patterns for sub-grains A and B is shown in Fig. 6.3, with the black circle indicating the position of the transmitted spot. The distance from the centre of the pattern to the transmitted spot, Z, for sub-grain B is measured and the misorientation is given by:

$$\tan\theta = \frac{Z}{L_{TEM}} \tag{6.8}$$

$$\text{or} \quad \theta = \tan^{-1}\left(\frac{Z}{L_{TEM}}\right) \tag{6.9}$$

where L_{TEM} is the camera length at which the diffraction patterns are taken.

Results

For the example, the distance Z in Fig. 6.3 is 0.011 m and the camera length is 1.2 m, giving a misorientation of 0.53°. Table 6.3 shows the results for all of the boundaries around sub-grains A and E in Fig. 6.2.

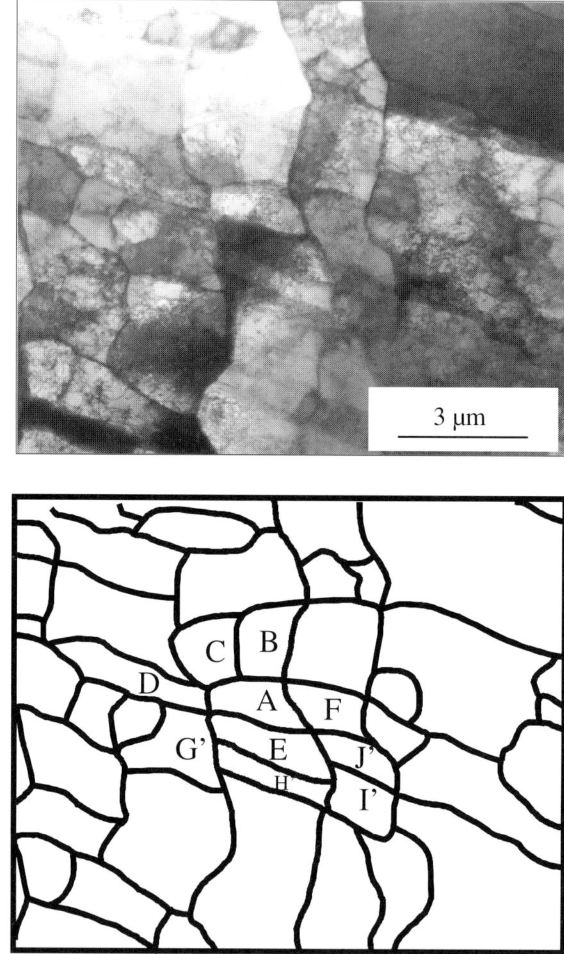

Fig. 6.2 Sub-grain structure in a hot deformed aluminium–magnesium alloy.

Table 6.3 Misorientations of the boundaries in Fig. 6.2 measured using CBED

Boundary	Misorientation (°)
A-B	0.53
A-C	3.54
A-D	2.69
A-E	4.02
A-F	3.11
E-G'	0.92
E-H'	1.20
E-I'	4.28
E-J'	1.26

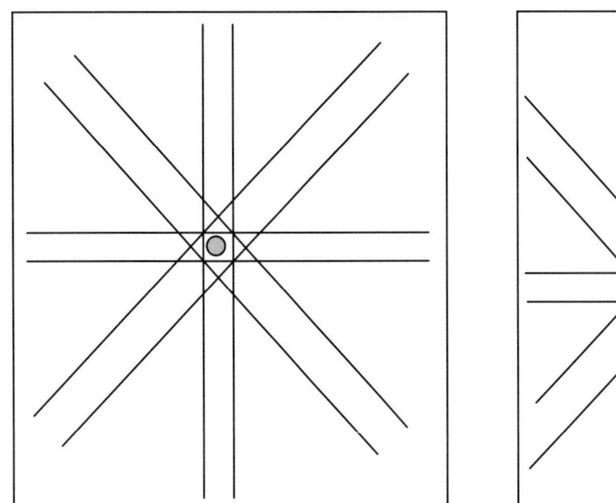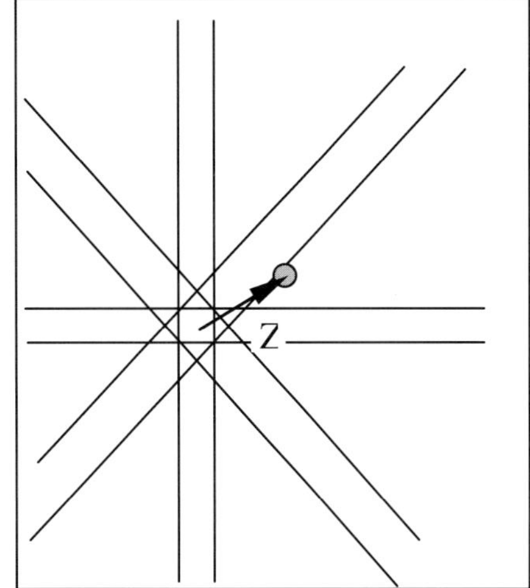

Fig. 6.3 A schematic of CBED Kikuchi patterns taken in sub-grains *A* and *B* in Fig. 6.2.

The mean and standard deviation for these measurements are:

$$\bar{\theta} = 2.39°$$

$$s = 1.43°$$

This gives a 95% confidence limit for $t_{95, 8}$ of:

$$\bar{\theta} = 2.4 \pm 1.1°$$

Comments

1. To cover a statistically significant number of boundaries using this method requires
    a large number of images to be taken from different foils and in different grains,
    and analysed. This can be time consuming and expensive, but may be unavoidable
    to obtain the required accuracy.
2. If samples of magnetic material are to be analysed, where tilting the sample can
    lead to difficulties, this method may be essential.

EXAMPLE 6.2.2 – FOIL TILTING

Method

As with the first method, a reference sub-grain is chosen, for Example A in Fig. 6.2.
Using CBED the sub-grain is then tilted to a known zone axis as before. For a more

Table 6.4 Misorientations of the boundaries in Fig. 6.2 measured by the foil tilting method.

Boundary	Misorientation ($°$)
A-B	0.53
A-C	3.62
A-D	2.71
A-E	4.26
A-F	3.38
E-G'	0.98
E-H'	1.31
E-I'	4.41
E-J'	1.32

accurate positioning the secondary diffraction patterns within the CBED discs (sometimes called dynamical fringes) can be used (Loretto (1994)). The tilts x and y are then noted. The beam is moved to a neighbouring sub-grain, for example B. This sub-grain is then tilted to the same zone axis as sub-grain A and again the tilts are noted. The misorientation between the sub-grains can then be calculated as:

$$\theta = \left(D_x^2 + D_y^2\right)^{1/2} \tag{6.10}$$

where D_x and D_y are the angular differences in the tilts between the two sub-grains of interest. The procedure is repeated for all the neighbouring sub-grains, C–F. The next reference sub-grain is then selected, for example E, and the procedure repeated for sub-grains G' to J' etc.

Results

For the sub-grains in Fig. 6.2 the position of the zone axis $B = [100]$ is at tilts of $x = 13.72°$ and $y = -10.6°$ for sub-grain A and $x = 13.42°$ and $y = -10.16°$ for sub-grain B. This gives a misorientation between these two sub-grains of:

$$\theta = \left(0.30^2 + 0.44^2\right)^{1/2}$$

giving $\theta = 0.533°$

Table 6.4 shows the results for all of the boundaries around sub-grains A and E in Fig. 6.2.

The mean and standard deviation for these measurements are

$$\bar{\theta} = 2.50°$$

$$s = 1.49°$$

This gives a 95% confidence limit for $t_{95,8}$ of

$$\bar{\theta} = 2.5 \pm 1.2°$$

Comments

1. Using this method, a large amount of data can be obtained without the need for images of diffraction patterns to be taken. The accuracy does, however, depend on the microscope operator and the goniometer. Tilting from sub-grain A to sub-grain B, for example, and then tilting back to sub-grain A and comparing the results can provide a check on the accuracy of the equipment.
2. The misorientation of the sub-grains will also vary from one grain to another due to the effects of grain orientation. So, regardless of the method used to measure the misorientations, care must be taken to carry out measurements on a large sample of material.

6.3 SUB-GRAIN SIZE

EXAMPLE 6.3.1 – EQUIAXED SUB-GRAINS

Method

The sub-grain size can be determined using the techniques discussed in Chapter 4. Taking the sub-structure in Fig. 6.4 and overlaying a circle of circumference 150 mm the number of intersections is 17 at a magnification of 5800×. The subgrain size for this circle is then

$$\delta = \frac{150}{5800 \times 17} = 1.521 \times 10^{-3} \, \text{mm}$$

Results

Nine further fields were counted and the results are shown in Table 6.5.
The mean sub-grain size and the standard deviation is

$$\bar{\delta} = 1.457 \times 10^{-3} \, \text{mm}$$

$$s = 0.213 \times 10^{-3} \, \text{mm}$$

This gives a 95% confidence limit for $t_{95,9}$ of

$$\bar{\delta} = 1.46 \pm 0.21 \, \mu\text{m}$$

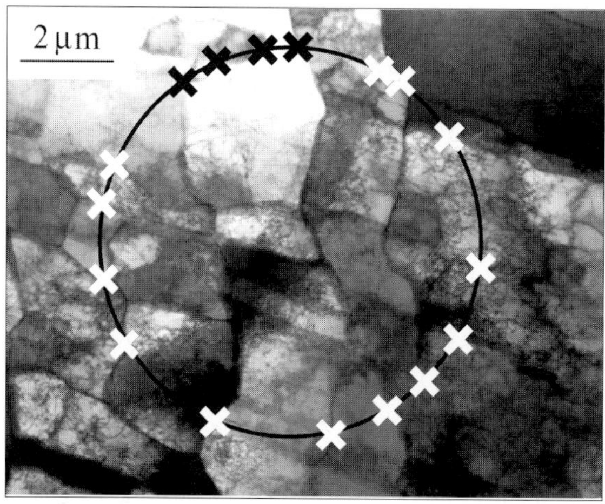

Fig. 6.4 Sub-grain size measured using linear intercepts with circles (150 mm circumference) on the micrograph shown in Fig. 6.2.

Table 6.5 Sub-grain size determined from 10 regions of thin foils.

Electron Micrograph	Number of Intersections	Sub-Grain Size (mm)
1	17	1.521×10^{-3}
2	18	1.437×10^{-3}
3	16	1.616×10^{-3}
4	20	1.293×10^{-3}
5	21	1.232×10^{-3}
6	17	1.521×10^{-3}
7	16	1.616×10^{-3}
8	23	1.124×10^{-3}
9	19	1.361×10^{-3}
10	14	1.847×10^{-3}

Comments

1. As when measuring other quantities using thin foils, it is important to take a large number of micrographs from different areas and different foils to obtain a statistically significant number of measurements.

Fig. 6.5 Electron-micrograph of a transverse section of Type 316 L stainless steel, rolled to a reduction of 50% at 900°C, ×6350.

EXAMPLE 6.3.2 – ELONGATED SUB-GRAINS

Most forming operations are carried out with the primary stress acting in a single direction. This can lead to elongated grain structures, as discussed in Chapter 4. It can, in the same way, lead to elongated sub-structures. This is further complicated by the formation of 'deformation' or 'micro' bands, which form at specific angles to the deformation direction. An example of an elongated sub-structure in hot rolled Type 316 L stainless steel is shown in Fig. 6.5.

It can be important to know the rolling/deformation direction in the foil. This can be achieved by careful alignment of the foil in the holder. The rolling direction is indicated in Fig. 6.5. The sub-grain size can then be determined by overlaying lines in the rolling and normal directions, as shown in Fig. 6.6.

Method

Lines of known length are overlaid on the electron micrograph of the sub-grains in the rolling and normal direction, as shown in Fig. 6.6. The number of intersections is then counted, as indicted by numbers at the end of the lines in the figure. As with all other measurements the procedure is repeated on a number of fields. A full set of results is shown in Table 6.6.

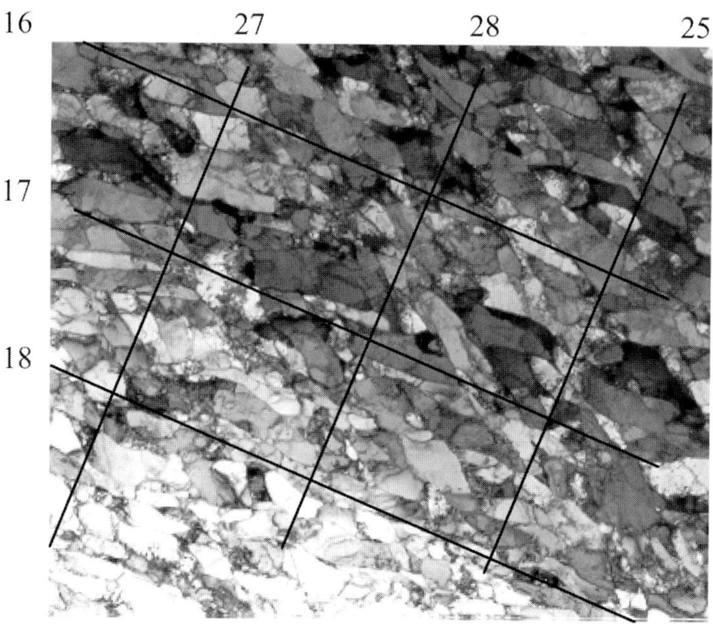

Fig. 6.6 Electron micrograph of hot rolled stainless steel, overlaid by lines for counting number of sub-grain boundaries.

Table 6.6 Results of linear intercept counts on Fig. 6.6 and three further areas.

Rolling Direction		Normal Direction	
No. Intercepts	**Line Length (mm)**	**No. Intercepts**	**Line Length (mm)**
16	87	27	69
17	87	28	69
18	87	25	69
17	90	31	75
15	90	29	75
14	90	34	75
13	82	20	65
12	82	23	65
15	82	25	65
16	86	32	68
12	86	24	68
15	86	28	68
Total = 180	Total = 1035	Total = 326	Total = 831

Using eqn (4.2) with a magnification of 6350× gives average sub-grain sizes of

$$\overline{\delta}_{RD} = \frac{1035}{6350 \times 180} = 9.06 \times 10^{-4} \text{ mm}$$

$$\overline{\delta}_{ND} = \frac{831}{6350 \times 326} = 4.01 \times 10^{-4} \text{ mm}$$

Using eqn (4.3) the standard errors are

$$S_{RD} = \frac{0.65}{\sqrt{180}} \times 9.06 \times 10^{-4} = 4.39 \times 10^{-5}$$

$$S_{ND} = \frac{0.65}{\sqrt{326}} \times 4.01 \times 10^{-4} = 1.44 \times 10^{-5}$$

Which gives sub boundary spacings of

$$\overline{\delta}_{RD} = 9.06 \times 10^{-4} \pm 4.39 \times 10^{-5} \text{ mm}$$

and

$$\overline{\delta}_{ND} = 4.01 \times 10^{-4} \pm 1.44 \times 10^{-5} \text{ mm}$$

or

$$\overline{\delta}_{RD} = 0.91 \pm 0.04 \text{ } \mu m$$

and

$$\overline{\delta}_{ND} = 0.40 \pm 0.01 \text{ } \mu m$$

Comments

1. The measurements of elongated sub-grain structures can be used to obtain other information about the sub-grains, as discussed for elongated grains in Chapter 4, e.g. an equivalent equiaxed sub-grain size

$$\overline{\delta} = \sqrt{\overline{\delta}_{RD} \times \overline{\delta}_{ND}} = \sqrt{0.91 \times 0.40} = 0.60 \mu m.$$

2. For elongated sub-grains the choice of orientations in which to make measurements may be selected to reflect local features, e.g. microbands, rather than the macroscopic deformation directions, which are relevant to grain shape measurements.

Appendix:
Statistical Methods for Data Analysis

There is a wide range of statistical methods available to handle all aspects of analysis of data obtained by random sampling. These are described in detail in many standard text books, for example Chatfield (1970), and this appendix only summarises the essential calculations required to obtain the confidence limits of measured mean values.

A.1 MEAN AND STANDARD DEVIATION

If measurements are made of many features in a sample, they will have variability about the mean value for the sample. It is usually assumed that the values (x) are distributed normally about the mean ($\bar{x}$) to give a frequency (f) distribution as illustrated in Fig. A.1. If n measurements are made, then

$$\bar{x} = \frac{\Sigma x}{n} \tag{A.1}$$

The width of the distribution is determined by the standard deviation(s), where

$$s^2 = \frac{\Sigma(x-\bar{x})^2}{n-1} \tag{A.2}$$

In a normal distribution, 67% of the measured values of x will fall within $\pm\,s$ of $\bar{x}$ and 95% will fall within $\pm\,2s$ of $\bar{x}$. Clearly, more consistent measurements of x lead to a narrower distribution and to a smaller value of s.

In quantitative metallography, measurements are frequently made by counting along traverse lines, e.g., for point fraction or for number per unit length. In these cases each traverse line provides a single measurement, no matter how many features are counted along the line, and n in eqns (A.1) and (A.2) is the number of traverse lines. A golden rule for the statistics of random sampling to be valid for quantitative metallography is that the same feature should never (or only very rarely) be measured twice, i.e. the traverse lines must be spaced more widely than the size of the features being counted.

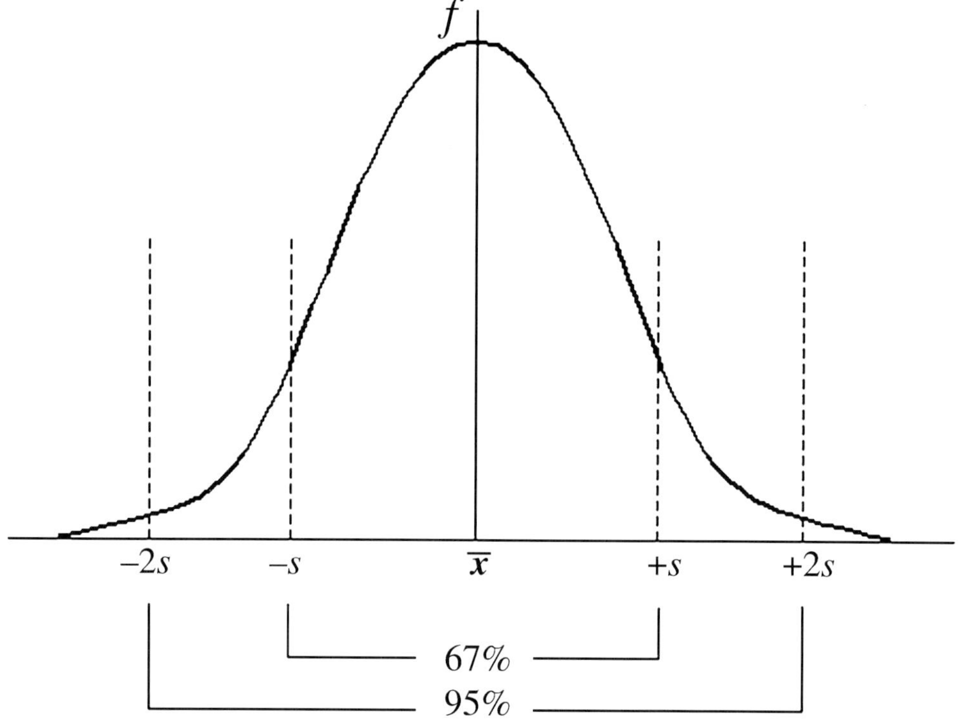

Fig. A.1 Normal frequency distribution curve of measurements of a parameter x in a sample.

A1.1 HISTOGRAMS

For determination of particle size, each individual particle is measured and n is the number of particles measured, which may be large (several hundred). In this case rather than measuring each particle accurately, the size may be put into a small size interval (δx) selected so that k (say, 10 to 15) size intervals covers the full range of particle sizes. The results may then be shown as a histogram, Fig. A.2. This should approximate to a normal distribution, shown, by the smooth curve in Fig. A.2. A histogram may be plotted as the number (n_i) in each size group (x_i), where i has values of 1 to k, or it may be plotted as frequency (f_i), as in Fig. A.2, where

$$f_i = \frac{n_i}{\Sigma n_i} = \frac{n_i}{n} \tag{A.3}$$

Data in the form of a histogram are analysed in the same way as before to obtain a mean

$$\overline{x} = \frac{\Sigma n_i x_i}{\Sigma n_i} = \frac{\Sigma n_i x_i}{n} \tag{A.4}$$

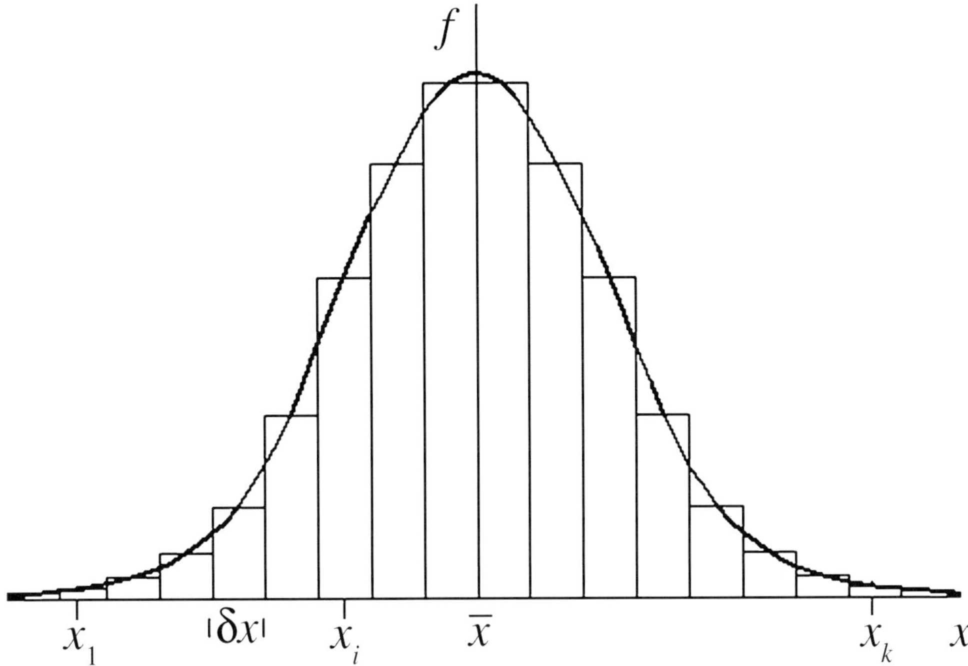

Fig. A.2 Histogram of measurements of a parameter x in a sample, grouped into size intervals δx.

where x_i is the mid-point of each size interval, i.e.,

$$x_i = (i - \tfrac{1}{2})(\delta x) \qquad (A.5)$$

The standard deviation becomes

$$s^2 = \frac{\Sigma n_i (x_i - \bar{x})^2}{\Sigma n - 1} = \frac{\Sigma n_i (x_i - \bar{x})^2}{n - 1} \qquad (A.6)$$

which may be rewritten as

$$s^2 = \frac{\Sigma n_i x_i^2 - n\bar{x}^2}{n - 1} \qquad (A.7)$$

This latter form is simpler to program on a calculator or set up on a spread sheet.

A.2 STANDARD ERROR AND CONFIDENCE LIMITS

The mean and standard deviation of a sample are usually determined to provide an estimate of the mean value of the population, i.e. in the bulk of the volume for quantitative

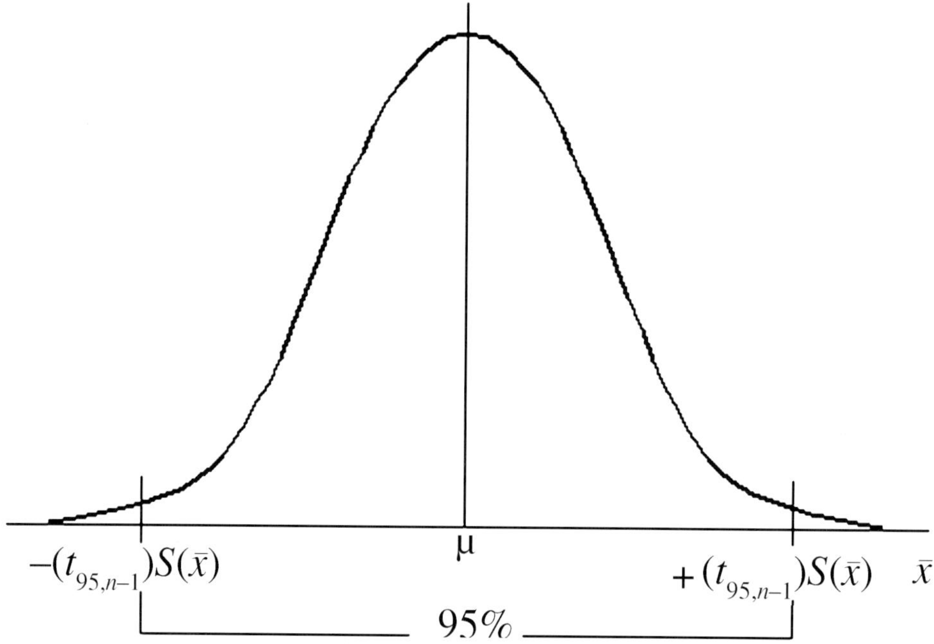

Fig. A3 *t*-distribution of sample means.

metallography. To do this, the variability of the sample mean must be determined. This is given by the standard error of the sample mean ($S(\bar{x})$), which is simply related to the standard deviation of the individual measurements as

$$S(\bar{x}) = \frac{s}{\sqrt{n}}$$

(A.8)

Clearly, the more measurements made in the determination of a sample mean, $\bar{x}$, then the smaller will be the variability of means of a number of different samples taken from the same population, which has a population mean (μ). In fact, the sample means ($\bar{x}$) will be distributed about μ in a *t*-distribution, Fig. A.3. The shape of this distribution changes with the number of measurements in each sample mean, so although the range of values in which 95% of the sample means fall is related to the standard error the multiplying constant ($t_{95,n-1}$) depends on the number of measurements in each sample, or on the number of degrees of freedom ($n-1$). This is illustrated in Fig. A4, from which it can be seen that when $n \geq 20$ the value of $t_{95,n-1} \approx 2$ because the *t*-distribution approximates to a normal distribution. For lower values of n, the appropriate (higher) value of $t_{95,n-1}$ must be used to obtain the 95% confidence interval. The range of these values can be seen in Fig. A4 and the exact values are given in Table A1.[*]

---

[*] It should be noted that columns in tables found in standard textbooks can be headed in different ways because when 95% of the values are inside the range, 5% are outside the range, with 2.5% above the upper limit. The correct column is the one in which the value of *t* approaches 2 as *n* increases (sometimes labelled $\alpha = 0.025$).

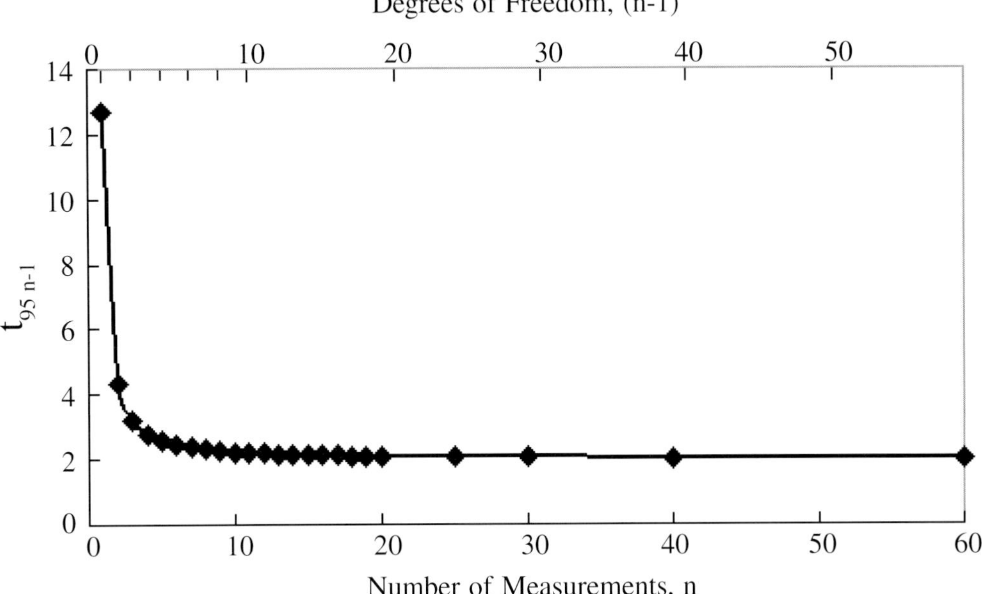

Fig. A4 Dependence of $t_{95,\,n-1}$ on the number of measurements made.

Table A1 Table of t values for 95% confidence limits as a function of the number of degrees of freedom $v = n - 1$.

v	t	v	t	v	t
1	12.706	10	2.228	19	2.093
2	4.303	11	2.201	20	2.086
3	3.182	12	2.179	25	2.060
4	2.776	13	2.160	30	2.042
5	2.571	14	2.145	40	2.021
6	2.447	15	2.131	60	2.000
7	2.365	16	2.120	120	1.98
8	2.306	17	2.110		
9	2.262	18	2.101		

When the 95% confidence interval is known, this can be applied to the single value of the mean $(\overline{x})$ determined from the experimental sample in order to define the range within which there is a 95% probability that the population mean value lies, i.e.

$$\mu = \bar{x} \pm (t_{95,\,n-1})\,S(\bar{x}) \tag{A.9}$$

Because it is generally the population mean that influences material properties etc., it is important that the confidence limits, as well as the sample mean, for microstructural parameters are reported. The limits also enable one to decide if two samples are from the same or from different populations, by applying the Students t-test. This test is outside the scope of the present summary, but is given in text books on statistics.

A.3 PROPAGATION OF ERRORS

When a variable (z) is a function of two or more independent variables, (say x and y), which must be measured experimentally,

$$\text{i.e. } z = F\,(x,\,y) \tag{A.10}$$

the standard error, or the 95% confidence interval, of z depends on the weighted average of the standard errors of x and y as

$$(S(\bar{z}))^2 = (S(\bar{x}))^2 \left(\frac{\partial z}{\partial x}\right)^2 + (S(\bar{y}))^2 \left(\frac{\partial z}{\partial y}\right)^2 + \ldots \ldots \tag{A.11}$$

For example, if

$$z = x^n\,y^m \tag{A.12}$$

$$(S(\bar{z}))^2 = (S(\bar{x}))^2\,(n\,x^{n-1}\,y^m)^2 + (S(\bar{y}))^2\,(m\,x^n\,y^{m-1})^2 \tag{A.13}$$

or, dividing by eqn (A.12), the relative standard error

$$\left(\frac{S(\bar{z})}{\bar{z}}\right)^2 = \left(\frac{S(\bar{x})}{\bar{x}}\right)^2 n^2 + \left(\frac{S(\bar{y})}{\bar{y}}\right)^2 m^2 \tag{A.14}$$

These relationships need to be used when measurements of grain size or colony size are made in duplex microstructures, Chapter 4.

References

ASTM Standards E112 (1999), American Society for Testing and Materials, Philadelphia, USA.

Ashby, M.F. and Ebeling, R. (1996), *Trans Met. Soc. AIME*, **236**, 1396.

Avrami, M. (1939), *J. Chem. Phys.*, **7**, 1103.

Cahn, J.W. and Hagel W. (1960), *Decomposition of Austenite by Diffusional Processes*, Z. D. Zackay and H. I. Aoronson eds, Interscience Pub., New York, 131.

Cahn, J.W. and Nutting, J. (1959), *Trans AIME*, **215**, 526.

Chatfield, C. (1970), *Statistics for Technology*, Penguin Books Ltd, Harmondsworth, Middlesex, England.

De Hoff, R.T. and Rhines, F.N. (1961), *Trans. AIME*, **221**, 975.

De Hoff, R.T. and Rhines, F.N. (1968), *Quantitative Metallography*, McGraw-Hill Book Co., New York.

Deiter, G.E. (1986), *Mechanical Metallurgy*, McGraw-Hill Book Company.

Delesse, A. (1848), *Ann. Mines*, **13**, 379.

Fullman, R.L. (1953), *Trans. AIM.E*, **197**, 4474.

Gladman, T. and Woodhead, J.H. (1960), *J. Iron Steel Inst.*, **194**, 189.

Gladman, T. (1963), *J. Iron Steel Inst.*, **201**, 1044.

Hilliard, J.E. (1962), *Trans. Met.Soc. AIME*, **224**, 906 .

Johnson, W.A. and Mehl, R.F. (1939), *Trans.AIME*, **135**, 416.

Kocks, U.F. (1966), *Phil. Mag.*, **13**, 541

Kolmogorov, A.N. (1937), *Izv. Akad. Nauk. USSR-Ser-Matemat.*, **1** (3), 355.

Loretto, M.H. (1994), *Electron Beam Analysis of Materials*, Chapman and Hall, London.

Mukherjee, R. Stumpf, W.E. and Sellars, C.M. (1968), *J. Mat. Sci.*, **3**, 127.

Orsetti-Rossi, P.L. and Sellars, C.M. (1997), *Acta Mater.*, **45**, 177–190.

Pereira da Silva, P.S.C. (1966), M. Met Thesis, The University of Sheffield.

Pickering, F.B. (1976), *The Basis of Quantitive Metallography*, The Institute of Metallurgists, London.

Pickering, F.B. (1983), *Physical Metallurgy and the Design of Steels*, Applied Science Publishers, London and New York.

Saltykov, S.A. (1952) " Stereometric Metallurgy", Ind. Ed., Metallurgizdat, Moscow.

Scheil, Z. (1935), *Z. Metallk.*, **27**, 199.

Schwartz, H.A. (1934), *Metals Alloys*, **5**,139.

Sellars, C.M. and Zhu, Q. (1999), *Proc. 20th Risø Int. Symp. on Mat. Sci.*, 167.

Shi, G. Atkinson, H.V., Sellars, C.M. and Anderson, C.W. (1999), *Ironmaking & Steelamaking*, **26** (4), 239.

Underwood, E.E. (1970), *Quantitative Stereology,* Addison-Wesley Publishing Co. Inc., Philippines.

Wells, M.A., Lloyd, D.J., Samarasekera, I.V., Brimacombe, J.K. and Hawbolt, E.B., (1998), *Metall. & Mater. Tran.,* **29B**, 611.

Woodhead, J. (1980), private communication.